CARL SAGAN

PALE BLUE DOT

A VISION OF THE
HUMAN FUTURE IN SPACE

Random House
New York

BELOW:

The Earth: a pale blue dot in a sunbeam.
Photographed by *Voyager 1*
from beyond the orbit of Neptune.

Originally published in hardcover by Random House, Inc., in 1994

LIBRARY OF CONGRESS CATALOGING-IN-PUBLICATION DATA
Sagan, Carl, 1934–
        Pale blue dot: a vision of the human future in space/Carl Sagan.
p. cm.
        Includes bibliographical references and index.
        ISBN 0-679-76486-0
        1. Outer space—Exploration—Popular works. I. Title.
QB500. 262.S24. 1994   919.9'04—dc20 94-18121

DESIGNED BY BETH TONDREAU DESIGN

Manufactured in the United States of America
9   8   7   6   5   4   3   2
First Paperback Edition

FOR SAM,
another wanderer.
May your generation see
wonders undreamt.

SPACECRAFT EXPLORATION

NOTABLE EARLY ACHIEVEMENTS

SOVIET UNION/RUSSIA

1957   First **artificial satellite of the Earth**
(*Sputnik 1*)

1957   First **animal in space**
(*Sputnik 2*)

1959   First **spacecraft to escape the Earth's gravity**
(*Luna 1*)

1959   First **artificial planet of the Sun**
(*Luna 1*)

1959   First **spacecraft to impact another world**
(*Luna 2* to the Moon)

1959   First **view of the far side of the moon**
(*Luna 3*)

1961   First **human in space**
(*Vostok 1*)

1961   First **human to orbit the Earth**
(*Vostok 1*)

1961   First **spacecraft to fly by other planets**
(*Venera 1* to Venus;
1962   *Mars 1* to Mars)

1963   First **woman in space**
(*Vostok 6*)

1964   First **multiperson space mission**
(*Voskhod 1*)

1965   First **space "walk"**
(*Voskhod 2*)

1966   First **spacecraft to enter the atmosphere of another planet**
(*Venera 3* to Venus)

1966   First **spacecraft to orbit another world**
(*Luna 10* to the Moon)

1966   First **successful soft landing on another world**
(*Luna 9* to the Moon)

1 9 5 0

1 9 6 0

OF THE SOLAR SYSTEM

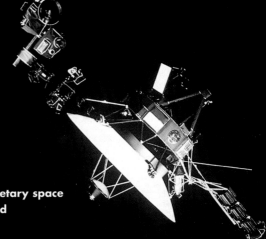

UNITED STATES

1958   First **scientific discovery in space
—Van Allen radiation belt**
(*Explorer 1*)

1959   First **television images of
the Earth from space**
(*Explorer 6*)

1962   First **scientific discovery in interplanetary space
—direct observation of the solar wind**
(*Mariner 2*)

1962   First **scientifically successful planetary mission**
(*Mariner 2* to Venus)

1962   First **astronomical observatory in space**
(*OSO-1*)

1968   First **manned orbit of another world**
(*Apollo 8* to the Moon)

1969   First **landing of humans on another world**
(*Apollo 11* to the Moon)

1969   First **samples returned to Earth from another world**
(*Apollo 11* to the Moon)

1971   First **manned roving vehicle on another world**
(*Apollo 15* to the Moon)

1971   First **spacecraft to orbit another planet**
(*Mariner 9* to Mars)

1974   First **dual–planet mission**
(*Mariner 10* to Venus and Mercury)

1976   First **successful Mars landing; first spacecraft
to search for life on another planet**
(*Viking 1*)

SPACECRAFT EXPLORATION

NOTABLE EARLY ACHIEVEMENTS

SOVIET UNION/RUSSIA

1970   First **robot mission to return a sample from another world**
(*Luna 16* to the Moon)

1970   First **roving vehicle on another world**
(*Luna 17* to the Moon)

1971   First **soft landing on another planet**
(*Mars 3* to Mars)

1972   First **scientifically successful landing on another planet**
(*Venera 8* to Venus)

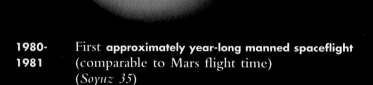

1980-1981   First **approximately year-long manned spaceflight**
(comparable to Mars flight time)
(*Soyuz 35*)

1983   First **full orbital radar mapping of another planet**
(*Venera 15* to Venus)

1985   First **balloon station deployed in the atmosphere of another planet**
(*Vega 1* to Venus)

1986   First **close cometary encounter**
(*Vega 1* to Halley's Comet)

1986   First **space station inhabited by rotating crews**
(*Mir*)

OF THE SOLAR SYSTEM

UNITED STATES

1973 First **flybys of Jupiter** (*Pioneer 10*),
1974 **Mercury** (*Mariner 10*),
1977 **Saturn** (*Pioneer 11*),

First **spacecraft to achieve escape velocity
from the Solar System**
(*Pioneers 10* and *11*,
launched in 1973 and 1974;
Voyagers 1 and *2*, 1977)

1981 First **manned reusable spacecraft**
(*STS-1*)

1980– First **satellite to be retrieved, repaired,
1984** **and redeployed in space**
(*Solar Maximum Mission*)

1985 First **distant cometary encounter**
(*International Cometary Explorer*
to Comet Giacobini–Zimmer)

1986 First **flybys of Uranus** (*Voyager 2*),
1989 **Neptune** (*Voyager 2*)

1992 First **detection of the heliopause**
(*Voyager*)

1992 First **encounter with a main-belt asteroid**
(*Galileo* to Gaspra)

1994 First **detection of a moon of an asteroid**
(*Galileo* to Ida)

CONTENTS

WANDERERS:
AN INTRODUCTION

But tell me, who *are* they, these wanderers . . . ?
—RAINER MARIA RILKE, "THE FIFTH ELEGY" (1923)

We were wanderers from the beginning. We knew every stand of tree for a hundred miles. When the fruits or nuts were ripe, we were there. We followed the herds in their annual migrations. We rejoiced in fresh meat. Through stealth, feint, ambush, and main-force assault, a few of us cooperating accomplished what many of us, each hunting alone, could not. We depended on one another. Making it on our own was as ludicrous to imagine as was settling down.

Working together, we protected our children from the lions

and the hyenas. We taught them the skills they would need. And the tools. Then, as now, technology was the key to our survival.

When the drought was prolonged, or when an unsettling chill lingered in the summer air, our group moved on—sometimes to unknown lands. We sought a better place. And when we couldn't get on with the others in our little nomadic band, we left to find a more friendly bunch somewhere else. We could always begin again.

For 99.9 percent of the time since our species came to be, we were hunters and foragers, wanderers on the savannahs and the steppes. There were no border guards then, no customs officials. The frontier was everywhere. We were bounded only by the Earth and the ocean and the sky—plus occasional grumpy neighbors.

When the climate was congenial, though, when the food was plentiful, we were willing to stay put. Unadventurous. Overweight. Careless. In the last ten thousand years—an instant in our long history—we've abandoned the nomadic life. We've domesticated the plants and animals. Why chase the food when you can make it come to you?

For all its material advantages, the sedentary life has left us edgy, unfulfilled. Even after 400 generations in villages and cities, we haven't forgotten. The open road still softly calls, like a nearly forgotten song of childhood. We invest far-off places with a certain romance. The appeal, I suspect, has been meticulously crafted by natural selection as an essential element in our survival. Long summers, mild winters, rich harvests, plentiful game—none of them lasts forever. It is beyond our powers to predict the future. Catastrophic events have a way of sneaking up on us, of catching us unaware. Your own life, or your band's, or even your species' might be owed to a restless few—drawn, by a craving they can hardly articulate or understand, to undiscovered lands and new worlds.

Herman Melville, in *Moby Dick,* spoke for wanderers in all epochs and meridians: "I am tormented with an everlasting itch for things remote. I love to sail forbidden seas . . ."

To the ancient Greeks and Romans, the known world comprised Europe and an attenuated Asia and Africa, all surrounded by an impassable World Ocean. Travelers might encounter inferior beings called barbarians or superior beings called gods. Every tree had its dryad, every district its legendary hero. But there were not very many gods, at least at first, perhaps only a few dozen. They lived on mountains, under the Earth, in the sea, or up there in the sky. They sent messages to people, intervened in human affairs, and interbred with us.

As time passed, as the human exploratory capacity hit its stride, there were surprises: Barbarians could be fully as clever as Greeks and Romans. Africa and Asia were larger than anyone had guessed. The World Ocean was not impassable. There were Antipodes.* Three new continents existed, had been settled by Asians in ages past, and the news had never reached Europe. Also the gods were disappointingly hard to find.

The first large-scale human migration from the Old World to the New happened during the last ice age, around 11,500 years ago, when the growing polar ice caps shallowed the oceans and made it possible to walk on dry land from Siberia to Alaska. A thousand years later, we were in Tierra del Fuego, the southern tip of South America. Long before Columbus, Indonesian argonauts in outrigger canoes explored the western Pacific; people from Borneo settled Madagascar; Egyptians and Libyans circumnavigated Africa; and a great fleet of oceangoing junks from Ming Dynasty China crisscrossed the Indian Ocean, established a base in Zanzibar, rounded the Cape of Good Hope, and entered the Atlantic Ocean. In the fifteenth through seventeenth centuries, European sailing ships discovered new continents (new, at any rate, to Europeans) and circumnavigated the planet. In the eighteenth and nineteenth centuries, American and Russian explorers, traders, and settlers raced west and east across two vast continents to the Pa-

---

* "As to the fable that there are Antipodes," wrote St. Augustine in the fifth century, "that is to say, men on the opposite side of the earth, where the sun rises when it sets to us, men who walk with their feet opposite ours, that is on no ground credible." Even if some unknown landmass is there, and not just ocean, "there was only one pair of original ancestors, and it is inconceivable that such distant regions should have been peopled by Adam's descendants."

cific. This zest to explore and exploit, however thoughtless its agents may have been, has clear survival value. It is not restricted to any one nation or ethnic group. It is an endowment that all members of the human species hold in common.

Since we first emerged, a few million years ago in East Africa, we have meandered our way around the planet. There are now people on every continent and the remotest islands, from pole to pole, from Mount Everest to the Dead Sea, on the ocean bottoms and even, occasionally, in residence 200 miles up—humans, like the gods of old, living in the sky.

These days there seems to be nowhere left to explore, at least on the land area of the Earth. Victims of their very success, the explorers now pretty much stay home.

Vast migrations of people—some voluntary, most not—have shaped the human condition. More of us flee from war, oppression, and famine today than at any other time in human history. As the Earth's climate changes in the coming decades, there are likely to be far greater numbers of environmental refugees. Better places will always call to us. Tides of people will continue to ebb and flow across the planet. But the lands we run to now have already been settled. Other people, often unsympathetic to our plight, are there before us.

LATE IN THE NINETEENTH CENTURY, Leib Gruber was growing up in Central Europe, in an obscure town in the immense, polyglot, ancient Austro-Hungarian Empire. His father sold fish when he could. But times were often hard. As a young man, the only honest employment Leib could find was carrying people across the nearby river Bug. The customer, male or female, would mount Leib's back; in his prized boots, the tools of his trade, he would wade out in a shallow stretch of the river and deliver his passenger to the opposite bank. Sometimes the water reached his waist. There were no bridges here, no ferryboats. Horses might have served the purpose, but they had other uses. That left Leib and a few other young men like him. *They* had no other uses. No other work was available. They would lounge about the riverbank, calling out their prices, boasting to potential customers about the superiority of their drayage. They hired themselves out like four-footed animals. My grandfather was a beast of burden.

I don't think that in all his young manhood Leib had ventured more than a hundred kilometers from his little hometown of Sassow. But then, in 1904, he suddenly ran away to the New World—to avoid a murder rap, according to one family legend. He left his young wife behind. How different from his tiny backwater hamlet the great German port cities must have seemed, how vast the ocean, how strange the lofty skyscrapers and endless hubbub of his new land. We know nothing of his crossing, but have found the ship's manifest for the journey undertaken later by his wife, Chaiya—joining Leib after he had saved enough to bring her over. She traveled in the cheapest class on the *Batavia,* a vessel of Hamburg registry. There's something heartbreakingly terse about the document: Can she read or write? No. Can she speak English? No. How much money does she have? I can imagine her vulnerability and her shame as she replies, "One dollar."

She disembarked in New York, was reunited with Leib, lived just long enough to give birth to my mother and her sister, and then died from "complications" of childbirth. In those few years in America, her name had sometimes been anglicized to Clara. A quarter century later, my mother named her own first-born, a son, after the mother she never knew.

OUR DISTANT ANCESTORS, watching the stars, noted five that did more than rise and set in stolid procession, as the so-called "fixed" stars did. These five had a curious and complex motion. Over the months they seemed to wander slowly among the stars. Sometimes they did loops. Today we call them planets, the Greek word for wanderers. It was, I imagine, a peculiarity our ancestors could relate to.

We know now that the planets are not stars, but other worlds, gravitationally lashed to the Sun. Just as the exploration of the Earth was being completed, we began to recognize it as one world among an uncounted multitude of others, circling the Sun or orbiting the other stars that make up the Milky Way Galaxy. Our planet and our solar system are surrounded by a new world ocean—the depths of space. It is no more impassable than the last.

Maybe it's a little early. Maybe the time is not quite yet. But those other worlds—promising untold opportunities—beckon.

In the last few decades, the United States and the former Soviet Union have accomplished something stunning and historic— the close-up examination of all those points of light, from Mercury to Saturn, that moved our ancestors to wonder and to science. Since the advent of successful interplanetary flight in 1962, our machines have flown by, orbited, or landed on more than seventy new worlds. We have wandered among the wanderers. We have found vast volcanic eminences that dwarf the highest mountain on Earth; ancient river valleys on two planets, enigmatically one too cold and the other too hot for running water; a giant planet with an interior of liquid metallic hydrogen into which a thousand Earths would fit; whole moons that have melted; a cloud-covered place with an atmosphere of corrosive acids, where even the high plateaus are above the melting point of lead; ancient surfaces on which a faithful record of the violent formation of the Solar System is engraved; refugee ice worlds from the transplutonian depths; exquisitely patterned ring systems, marking the subtle harmonies of gravity; and a world surrounded by clouds of complex organic molecules like those that in the earliest history of our planet led to the origin of life. Silently, they orbit the Sun, waiting.

We have uncovered wonders undreamt by our ancestors who first speculated on the nature of those wandering lights in the night sky. We have probed the origins of our planet and ourselves. By discovering what else is possible, by coming face to face with alternative fates of worlds more or less like our own, we have begun to better understand the Earth. Every one of these worlds is lovely and instructive. But, so far as we know, they are also, every one of them, desolate and barren. Out there, there are no "better places." So far, at least.

During the *Viking* robotic mission, beginning in July 1976, in a certain sense I spent a year on Mars. I examined the boulders and sand dunes, the sky red even at high noon, the ancient river valleys, the soaring volcanic mountains, the fierce wind erosion, the laminated polar terrain, the two dark potato-shaped moons. But there was no life—not a cricket or a blade of grass, or even, so far as we can tell for sure, a microbe. These worlds have not been graced, as ours has, by life. Life is a comparative rarity. You can survey dozens of worlds and find that on only one of them does life arise and evolve and persist.

Having in all their lives till then crossed nothing wider than a river, Leib and Chaiya graduated to crossing oceans. They had one great advantage: On the other side of the waters there would be—invested with outlandish customs, it is true—other human beings speaking their language and sharing at least some of their values, even people to whom they were closely related.

In our time we've crossed the Solar System and sent four ships to the stars. Neptune lies a million times farther from Earth than New York City is from the banks of the Bug. But there are no distant relatives, no humans, and apparently no life waiting for us on those other worlds. No letters conveyed by recent emigrés help us to understand the new land—only digital data transmitted at the speed of light by unfeeling, precise robot emissaries. They tell us that these new worlds are not much like home. But we continue to search for inhabitants. We can't help it. Life looks for life.

No one on Earth, not the richest among us, can afford the passage; so we can't pick up and leave for Mars or Titan on a whim, or because we're bored, or out of work, or drafted into the army, or oppressed, or because, justly or unjustly, we've been accused of a crime. There does not seem to be sufficient short-term profit to motivate private industry. If we humans ever go to these worlds, then, it will be because a nation or a consortium of them believes it to be to its advantage—or to the advantage of the human species. Just now, there are a great many matters pressing in on us that compete for the money it takes to send people to other worlds.

That's what this book is about: other worlds, what awaits us on them, what they tell us about ourselves, and—given the urgent problems our species now faces—whether it makes sense to go. Should we solve those problems first? Or are they a reason for going?

This book is, in many ways, optimistic about the human prospect. The earliest chapters may at first sight seem to revel overmuch in our imperfections. But they lay an essential spiritual and logical foundation for the development of my argument.

I have tried to present more than one facet of an issue. There will be places where I seem to be arguing with myself. I am. Seeing some merit to more than one side, I often argue with myself. I hope by the last chapter it will be clear where I come out.

The plan of the book is roughly this: We first examine the widespread claims made over all of human history that our world and our species are unique, and even central to the workings and purpose of the Cosmos. We venture through the Solar System in the footsteps of the latest voyages of exploration and discovery, and then assess the reasons commonly offered for sending humans into space. In the last and most speculative part of the book, I trace how I imagine that our long-term future in space will work itself out.

Pale Blue Dot is about a new recognition, still slowly overtaking us, of our coordinates, our place in the Universe—and how, even if the call of the open road is muted in our time, a central element of the human future lies far beyond the Earth.

CAPTION FOR PAGE x: The worlds of the Solar System as known towards the end of the age of preliminary spacecraft reconnaissance. The terrestrial planets, except for Mercury and the Galilean satellites of Jupiter, are shown in three different meridians. Some of the moons of Saturn and Uranus are shown in two different meridians. No detail is shown for Titan, because we know almost nothing about its surface. Parts of some worlds—for example, Rhea, Callisto, and Mercury—show very little detail, because these regions have never been visited close-up by interplanetary spacecraft. Pluto and Charon show detail as inferred from Earthbound occultation observations. Many of the small moons in the outer Solar System are not shown. The worlds are shown here to scale, except as indicated. (Mimas, for example, is shown three times the size it would really be compared, say, to the Earth.) The overwhelming majority of the data on which this figure is based were obtained from spacecraft launched by the National Aeronautics and Space Administration (NASA). The Venus data comes in part from spacecraft of the Soviet Union, and the information on Halley's Comet from a spacecraft of the European Space Agency. Courtesy NASA and USGS. A wall chart of this illustration is for sale by the U.S. Geological Survey, Map Distribution, Box 25286, Federal Center, Denver, CO 80225.

□
YOU ARE HERE

YOU ARE HERE

The entire Earth is but a point, and the place of
our own habitation but a minute corner of it.
—MARCUS AURELIUS, ROMAN EMPEROR,
MEDITATIONS, BOOK 4 (CA. 170)

As the astronomers unanimously teach, the circuit of the whole earth,
which to us seems endless, compared with the greatness of the universe
has the likeness of a mere tiny point.
—AMMIANUS MARCELLINUS (CA. 330–395),
THE LAST MAJOR ROMAN HISTORIAN,
IN *THE CHRONICLE OF EVENTS*

The spacecraft was a long way from home, beyond the orbit of the outermost planet and high above the ecliptic plane—which is an imaginary flat surface that we can think of as something like a racetrack in which the orbits of the planets are mainly confined. The ship was speeding away from the Sun at 40,000 miles per hour. But in early February of 1990, it was overtaken by an urgent message from Earth.

Obediently, it turned its cameras back toward the now-distant planets. Slewing its scan platform from one spot in the sky to an-

OPPOSITE: The position of the Earth and Sun (and many of the stars in our night sky) as seen from a vantage point outside the Milky Way. How this scene fits into the structure of our galaxy can be made out by comparing it with pages 380–81. Painting by Jon Lomberg.

other, it snapped 60 pictures and stored them in digital form on its tape recorder. Then, slowly, in March, April, and May, it radioed the data back to Earth. Each image was composed of 640,000 individual picture elements ("pixels"), like the dots in a newspaper wirephoto or a pointillist painting. The spacecraft was 3.7 billion miles away from Earth, so far away that it took each pixel $5^{1}/_{2}$ hours, traveling at the speed of light, to reach us. The pictures might have been returned earlier, but the big radio telescopes in California, Spain, and Australia that receive these whispers from the edge of the Solar System had responsibilities to other ships that ply the sea of space—among them, *Magellan,* bound for Venus, and *Galileo* on its tortuous passage to Jupiter.

Voyager 1 was so high above the ecliptic plane because, in 1981, it had made a close pass by Titan, the giant moon of Saturn. Its sister ship, *Voyager 2,* was dispatched on a different trajectory, within the ecliptic plane, and so she was able to perform her celebrated explorations of Uranus and Neptune. The two *Voyager* robots have explored four planets and nearly sixty moons. They are triumphs of human engineering and one of the glories of the American space program. They will be in the history books when much else about our time is forgotten.

The *Voyagers* were guaranteed to work only until the Saturn encounter. I thought it might be a good idea, just after Saturn, to have them take one last glance homeward. From Saturn, I knew, the Earth would appear too small for *Voyager* to make out any detail. Our planet would be just a point of light, a lonely pixel, hardly distinguishable from the many other points of light *Voyager* could see, nearby planets and far-off suns. But precisely because of the obscurity of our world thus revealed, such a picture might be worth having.

Mariners had painstakingly mapped the coastlines of the continents. Geographers had translated these findings into charts and globes. Photographs of tiny patches of the Earth had been obtained first by balloons and aircraft, then by rockets in brief ballistic flight, and at last by orbiting spacecraft—giving a perspective like the one you achieve by positioning your eyeball about an inch above a large globe. While almost everyone is taught that the Earth

The whole Earth photographed on the *Apollo 17* mission. Courtesy NASA.

is a sphere with all of us somehow glued to it by gravity, the reality of our circumstance did not really begin to sink in until the famous frame-filling *Apollo* photograph of the whole Earth—the one taken by the *Apollo 17* astronauts on the last journey of humans to the Moon.

It has become a kind of icon of our age. There's Antarctica at what Americans and Europeans so readily regard as the bottom, and then all of Africa stretching up above it: You can see Ethiopia, Tanzania, and Kenya, where the earliest humans lived. At top right are Saudi Arabia and what Europeans call the Near East. Just barely peeking out at the top is the Mediterranean Sea, around which so much of our global civilization emerged. You can make out the blue of the ocean, the yellow-red of the Sahara and the Arabian desert, the brown-green of forest and grassland.

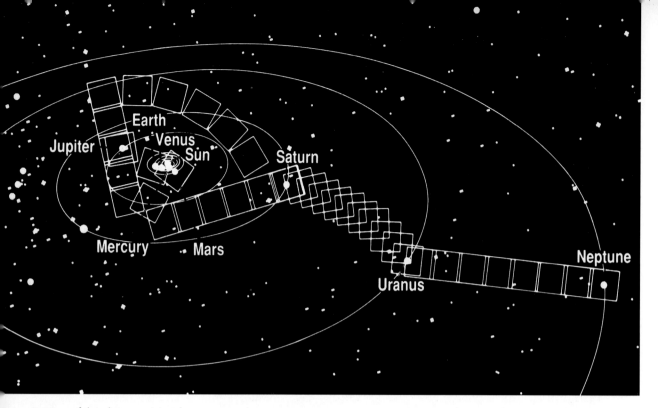

Jupiter · Earth · Venus · Sun · Saturn · Mercury · Mars · Uranus · Neptune

Position of the planets against the background of more distant stars at the moment *Voyager 1* took its family portrait of the Solar System. The Sun and the inner planets out to Mars are tightly clustered left of center. The outside four orbits are of Jupiter, Saturn, Uranus, and Neptune. The squares show the positions of the individual spacecraft imaging frames as laid down on the sky. This view is possible only because *Voyager 1* was high above the ecliptic plane in which the planets revolve about the Sun. The Earth is seen as an individual picture element, but Jupiter (and Saturn with its rings) are larger than a single dot. Courtesy JPL/NASA.

And yet there is no sign of humans in this picture, not our re-working of the Earth's surface, not our machines, not ourselves: We are too small and our statecraft is too feeble to be seen by a space-craft between the Earth and the Moon. From this vantage point, our obsession with nationalism is nowhere in evidence. The *Apollo* pictures of the whole Earth conveyed to multitudes something well known to astronomers: On the scale of worlds—to say noth-ing of stars or galaxies—humans are inconsequential, a thin film of life on an obscure and solitary lump of rock and metal.

It seemed to me that another picture of the Earth, this one taken from a hundred thousand times farther away, might help in the continuing process of revealing to ourselves our true circum-stance and condition. It had been well understood by the scientists and philosophers of classical antiquity that the Earth was a mere point in a vast encompassing Cosmos, but no one had ever *seen* it as such. Here was our first chance (and perhaps also our last for decades to come).

Many in NASA's *Voyager* Project were supportive. But from the outer Solar System the Earth lies very near the Sun, like a moth enthralled around a flame. Did we want to aim the camera so close to the Sun as to risk burning out the spacecraft's vidicon system? Wouldn't it be better to delay until all the scientific images—from Uranus and Neptune, if the spacecraft lasted that long—were taken?

And so we waited, and a good thing too—from 1981 at Saturn, to 1986 at Uranus, to 1989, when both spacecraft had passed the orbits of Neptune and Pluto. At last the time came. But there were a few instrumental calibrations that needed to be done first, and we waited a little longer. Although the spacecraft were in the right spots, the instruments were still working beautifully, and there were no other pictures to take, a few project personnel opposed it. It wasn't science, they said. Then we discovered that the technicians who devise and transmit the radio commands to *Voyager* were, in a cash-strapped NASA, to be laid off immediately or transferred to other jobs. If the picture were to be taken, it had to be done right then. At the last minute—actually, in the midst of the *Voyager 2* encounter with Neptune—the then NASA Administrator, Rear Admiral Richard Truly, stepped in and made sure that these images were obtained. The space scientists Candy Hansen of NASA's Jet Propulsion Laboratory (JPL) and Carolyn Porco of the University of Arizona designed the command sequence and calculated the camera exposure times.

So here they are—a mosaic of squares laid down on top of the planets and a background smattering of more distant stars. We were able to photograph not only the Earth, but also five other of the Sun's nine known planets. Mercury, the innermost, was lost in the glare of the Sun, and Mars and Pluto were too small, too dimly lit, and/or too far away. Uranus and Neptune are so dim that to record their presence required long exposures; accordingly, their images were smeared because of spacecraft motion. This is how the planets would look to an alien spaceship approaching the Solar System after a long interstellar voyage.

From this distance the planets seem only points of light, smeared or unsmeared—even through the high-resolution telescope aboard *Voyager*. They are like the planets seen with the naked eye from the surface of the Earth—luminous dots, brighter than most of the stars. Over a period of months the Earth, like the other planets, would seem to move among the stars. You cannot tell merely by looking at one of these dots what it's like, what's on

BELOW AND ON THE FOLLOWING PAGE: Six of the nine planets, photographed on February 14, 1990 from beyond the orbits of Neptune and Pluto by *Voyager 1*. Courtesy JPL and NASA.

VENUS

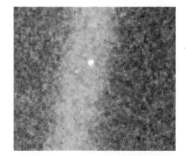

EARTH

JUPITER

SATURN

URANUS

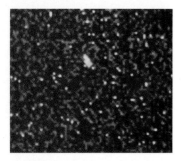

NEPTUNE

it, what its past has been, and whether, in this particular epoch, anyone lives there.

Because of the reflection of sunlight off the spacecraft, the Earth seems to be sitting in a beam of light, as if there were some special significance to this small world. But it's just an accident of geometry and optics. The Sun emits its radiation equitably in all directions. Had the picture been taken a little earlier or a little later, there would have been no sunbeam highlighting the Earth.

And why that cerulean color? The blue comes partly from the sea, partly from the sky. While water in a glass is transparent, it absorbs slightly more red light than blue. If you have tens of meters of the stuff or more, the red light is absorbed out and what gets re-flected back to space is mainly blue. In the same way, a short line of sight through air seems perfectly transparent. Nevertheless—something Leonardo da Vinci excelled at portraying—the more distant the object, the bluer it seems. Why? Because the air scatters blue light around much better than it does red. So the bluish cast of this dot comes from its thick but transparent atmosphere and its deep oceans of liquid water. And the white? The Earth on an aver-age day is about half covered with white water clouds.

We can explain the wan blueness of this little world because we know it well. Whether an alien scientist newly arrived at the outskirts of our solar system could reliably deduce oceans and clouds and a thickish atmosphere is less certain. Neptune, for in-stance, is blue, but chiefly for different reasons. From this distant vantage point, the Earth might not seem of any particular interest.

But for us, it's different. Look again at that dot. That's here. That's home. That's us. On it everyone you love, everyone you know, everyone you ever heard of, every human being who ever was, lived out their lives. The aggregate of our joy and suffering, thousands of confident religions, ideologies, and economic doc-trines, every hunter and forager, every hero and coward, every cre-ator and destroyer of civilization, every king and peasant, every young couple in love, every mother and father, hopeful child, in-ventor and explorer, every teacher of morals, every corrupt politi-cian, every "superstar," every "supreme leader," every saint and sinner in the history of our species lived there—on a mote of dust suspended in a sunbeam.

The Earth is a very small stage in a vast cosmic arena. Think of the rivers of blood spilled by all those generals and emperors so that, in glory and triumph, they could become the momentary masters of a fraction of a dot. Think of the endless cruelties visited by the inhabitants of one corner of this pixel on the scarcely distinguishable inhabitants of some other corner, how frequent their misunderstandings, how eager they are to kill one another, how fervent their hatreds.

Our posturings, our imagined self-importance, the delusion that we have some privileged position in the Universe, are challenged by this point of pale light. Our planet is a lonely speck in the great enveloping cosmic dark. In our obscurity, in all this vastness, there is no hint that help will come from elsewhere to save us from ourselves.

The Earth is the only world known so far to harbor life. There is nowhere else, at least in the near future, to which our species could migrate. Visit, yes. Settle, not yet. Like it or not, for the moment the Earth is where we make our stand.

It has been said that astronomy is a humbling and character-building experience. There is perhaps no better demonstration of the folly of human conceits than this distant image of our tiny world. To me, it underscores our responsibility to deal more kindly with one another, and to preserve and cherish the pale blue dot, the only home we've ever known.

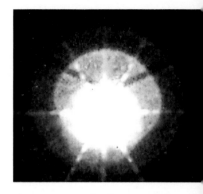

The Sun as seen by the *Voyager 1* wide-angle camera through its darkest filter and with the shortest possible exposure (0.005 seconds). From beyond the outermost planet, the Sun is only about 1/40 its size as seen from Earth. But it is still almost 8 million times brighter than Sirius, the brightest star in our sky. Courtesy JPL/NASA.

ABERRATIONS OF LIGHT

If man were taken away from the world, the rest would seem to be all astray,
without aim or purpose . . . and to be leading to nothing.
—FRANCIS BACON, *WISDOM OF THE ANCIENTS* (1619)

Ann Druyan suggests an experiment: Look back again at the pale blue dot of the preceding chapter. Take a good long look at it. Stare at the dot for any length of time and then try to convince yourself that God created the whole Universe for one of the 10 million or so species of life that inhabit that speck of dust. Now take it a step further: Imagine that everything was made just for a single shade of that species, or gender, or ethnic or religious subdivision. If this doesn't strike you as unlikely, pick another dot. Imagine *it* to be inhabited by a different form of intelligent life. They, too, cherish the notion of a God who has created everything for their benefit. How seriously do you take *their* claim?

OPPOSITE: The stars rise and set around us, encouraging the belief that the Earth is at the center of the Universe. In this time exposure, the center of the Milky Way in the constellation Sagittarius is seen. Every star is a sun. There are about 400 billion of them in the Milky Way. Photographed by Frank Zullo, Superstition Mountains, Arizona. Copyright ©1987 by Frank Zullo.

The nebula surrounding the dying star Eta Carinae. The nebula is formed from a succession of violent outbursts by the star. (The last outburst was observed in 1841, when despite its distance—over 10,000 light-years from Earth—Eta Carinae briefly became the second brightest star in our sky.) Were we as far from Eta Carinae as we are in fact from the Sun, it would appear 4 million times brighter than the Sun. The surface of the Earth would melt—rocks, mountains, and all. Courtesy Anglo-Australian Observatory. Photograph by David Malin.

OPPOSITE: A close-up of Eta Carinae before civilization arose on Earth—as seen by the Hubble Space Telescope. Two vast clouds of starstuff have been expelled, one (to the left) traveling approximately in our direction, and the other (upper right) traveling away. Scenes of cosmic violence are a staple of modern astronomy. Courtesy J. Hester, Arizona State University and NASA.

"SEE THAT STAR?"

"You mean the bright red one?" his daughter asks in return.

"Yes. You know, it might not be there anymore. It might be gone by now—exploded or something. Its light is still crossing space, just reaching our eyes now. But we don't see it as it is. We see it as it was."

Many people experience a stirring sense of wonder when they first confront this simple truth. Why? Why should it be so

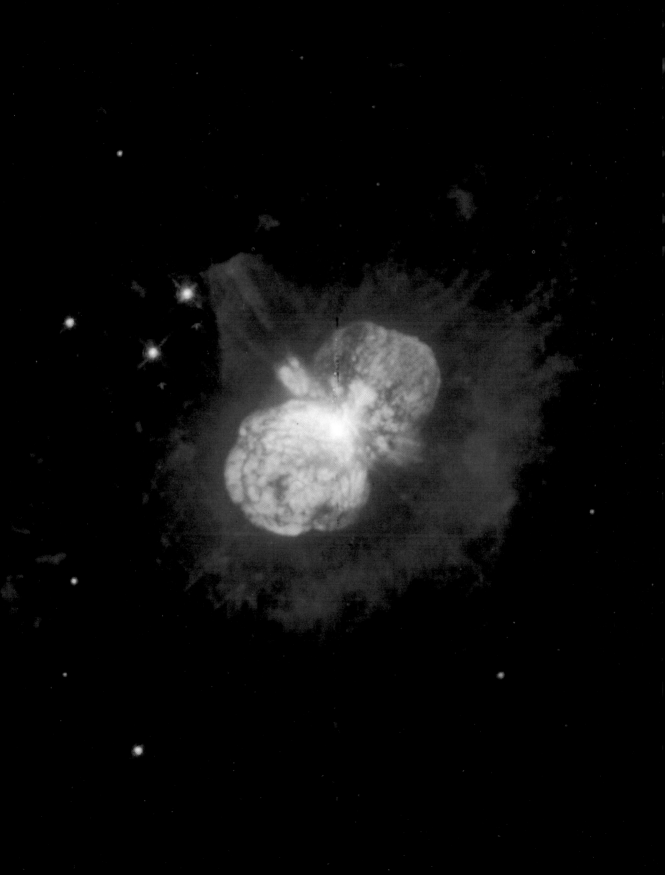

compelling? On our little world light travels, for all practical purposes, instantaneously. If a lightbulb is glowing, then of course it's
physically where we see it, shining away. We reach out our hand
and touch it: It's there all right, and unpleasantly hot. If the fila-
ment fails, then the light goes out. We don't see it in the same
place, glowing, illuminating the room years after the bulb breaks
and it's removed from its socket. The very notion seems nonsensi-
cal. But if we're far enough away, an entire sun can go out and
we'll continue to see it shining brightly; we won't learn of its
death, it may be, for ages to come—in fact, for how long it takes
light, which travels fast but not infinitely fast, to cross the intervening vastness.

The immense distances to the stars and the galaxies mean that
we see everything in space in the past—some as they were before
the Earth came to be. Telescopes are time machines. Long ago,
when an early galaxy began to pour light out into the surrounding
darkness, no witness could have known that billions of years later
some remote clumps of rock and metal, ice and organic molecules
would fall together to make a place called Earth; or that life would
arise and thinking beings evolve who would one day capture a little of that galactic light, and try to puzzle out what had sent it on
its way.

And after the Earth dies, some 5 billion years from now, after
it is burned to a crisp or even swallowed by the Sun, there will be
other worlds and stars and galaxies coming into being—and they
will know nothing of a place once called Earth.

IT ALMOST NEVER FEELS like prejudice. Instead, it seems fitting and
just—the idea that, because of an accident of birth, *our* group
(whichever one it is) should have a central position in the social
universe. Among Pharaonic princelings and Plantagenet pre-
tenders, children of robber barons and Central Committee bu-
reaucrats, street gangs and conquerors of nations, members of
confident majorities, obscure sects, and reviled minorities, this self-
serving attitude seems as natural as breathing. It draws sustenance
from the same psychic wellsprings as sexism, racism, nationalism,
and the other deadly chauvinisms that plague our species. Uncom-

mon strength of character is needed to resist the blandishments of those who assure us that we have an obvious, even God-given, superiority over our fellows. The more precarious our self-esteem, the greater our vulnerability to such appeals.

Since scientists are people, it is not surprising that comparable pretensions have insinuated themselves into the scientific worldview. Indeed, many of the central debates in the history of science seem to be, in part at least, contests over whether humans are special. Almost always, the going-in assumption is that we *are* special. After the premise is closely examined, though, it turns out—in dishearteningly many cases—that we are not.

Our ancestors lived out of doors. They were as familiar with the night sky as most of us are with our favorite television programs. The Sun, the Moon, the stars, and the planets all rose in the east and set in the west, traversing the sky overhead in the interim. The motion of the heavenly bodies was not merely a diversion, eliciting a reverential nod and grunt; it was the only way to tell the time of day and the seasons. For hunters and gatherers, as well as for agricultural peoples, knowing about the sky was a matter of life and death.

How lucky for us that the Sun, the Moon, the planets, and the stars are part of some elegantly configured cosmic clockwork! It seemed to be no accident. They were put here for a purpose, for our benefit. Who else makes use of them? What else are they good for?

And if the lights in the sky rise and set around us, isn't it evident that we're at the center of the Universe? These celestial bodies—so clearly suffused with unearthly powers, especially the Sun on which we depend for light and heat—circle us like courtiers fawning on a king. Even if we had not already guessed, the most elementary examination of the heavens reveals that we *are* special. The Universe seems designed for human beings. It's difficult to contemplate these circumstances without experiencing stirrings of pride and reassurance. The entire Universe, made for us! We must really be something.

This satisfying demonstration of our importance, buttressed by daily observations of the heavens, made the geocentrist conceit

A token of the astonishing
richness of the Milky Way. In the
original image there are perhaps
ten thousand stars in this picture,
a large number, but constituting
less than one ten-millionth of the
number of stars in the Galaxy.
The nebula glowing in hydrogen
gas at top left is M17.
Courtesy ROE/Anglo-Australian
Observatory. Photograph by
David Malin.

a transcultural truth—taught in the schools, built into the language, part and parcel of great literature and sacred scripture. Dissenters were discouraged, sometimes with torture and death. It is no wonder that for the vast bulk of human history, no one questioned it.

It was doubtless the view of our foraging and hunting ancestors. The great astronomer of antiquity, Claudius Ptolemaeus (Ptolemy), in the second century knew that the Earth was a sphere, knew that its size was "a point" compared to the distance of the stars, and taught that it lay "right in the middle of the heavens." Aristotle, Plato, St. Augustine, St. Thomas Aquinas, and almost all the great philosophers and scientists of all cultures over the 3,000 years ending in the seventeenth century bought into this delusion. Some busied themselves figuring out how the Sun, the Moon, the stars, and the planets could be cunningly attached to perfectly transparent, crystalline spheres—the big spheres, of course, centered on the Earth—that would explain the complex motions of the celestial bodies so meticulously chronicled by generations of astronomers. And they succeeded: With later modifications, the geocentric hypothesis adequately accounted for the facts of planetary motion as known in the second century, and in the sixteenth.

From there it was only a slight extrapolation to an even more grandiose claim—that the "perfection" of the world would be incomplete without humans, as Plato asserted in the *Timaeus.* "Man . . . is all," the poet and cleric John Donne wrote in 1625. "He is not a piece of the world, but the world itself; and next to the glory of God, the reason why there is a world."

And yet—never mind how many kings, popes, philosophers, scientists, and poets insisted on the contrary—the Earth through those millennia stubbornly persisted in orbiting the Sun. You might imagine an uncharitable extraterrestrial observer looking down on our species over all that time—with us excitedly chattering, "The Universe created for us! We're at the center! Everything pays homage to us!"—and concluding that our pretensions are amusing, our aspirations pathetic, that this must be the planet of the idiots.

But such a judgment is too harsh. We did the best we could. There was an unlucky coincidence between everyday appearances and our secret hopes. We tend not to be especially critical when presented with evidence that seems to confirm our prejudices. And there was little countervailing evidence.

In muted counterpoint, a few dissenting voices, counseling humility and perspective, could be heard down through the centuries. At the dawn of science, the atomist philosophers of ancient Greece and Rome—those who first suggested that matter is made of atoms—Democritus, Epicurus, and their followers (and Lucretius, the first popularizer of science) scandalously proposed many worlds and many alien life forms, all made of the same kinds of atoms as we. They offered for our consideration infinities in space and time. But in the prevailing canons of the West, secular and sacerdotal, pagan and Christian, atomist ideas were reviled. Instead, the heavens were not at all like our world. They were unalterable and "perfect." The Earth was mutable and "corrupt." The Roman statesman and philosopher Cicero summarized the common view: "In the heavens ... there is nothing of chance or hazard, no error, no frustration, but absolute order, accuracy, calculation and regularity."

Philosophy and religion cautioned that the gods (or God) were far more powerful than we, jealous of their prerogatives and quick to mete out justice for insufferable arrogance. At the same time, these disciplines had not a clue that their own teaching of how the Universe is ordered was a conceit and a delusion.

Philosophy and religion presented mere opinion—opinion that might be overturned by observation and experiment—as certainty. This worried them not at all. That some of their deeply held beliefs might turn out to be mistakes was a possibility hardly considered. Doctrinal humility was to be practiced by others. Their own teachings were inerrant and infallible. In truth, they had better reason to be humble than they knew.

BEGINNING WITH COPERNICUS in the middle sixteenth century, the issue was formally joined. The picture of the Sun rather than the Earth at the center of the Universe was understood to be dangerous. Obligingly, many scholars were quick to assure the religious hierarchy that this newfangled hypothesis represented no serious challenge to conventional wisdom. In a kind of split-brain compromise, the Sun-centered system was treated as a mere computational convenience, not an astronomical reality—that is, the Earth was *really* at the center of the Universe, as everybody knew; but if you wished to predict where Jupiter would be on the second Tuesday of November the year after next, you were permitted to pretend that the Sun was at the center. Then you could calculate away and not affront the Authorities.*

"This has no danger in it," wrote Robert Cardinal Bellarmine, the foremost Vatican theologian in the early seventeenth century,

> and suffices for the mathematicians. But, to affirm that the Sun is really fixed in the center of the heavens and that the Earth revolves very swiftly around the Sun is a dangerous thing, not only irritating the theologians and philosophers, but injuring our holy faith and making the sacred scripture false.

"Freedom of belief is pernicious," Bellarmine wrote on another occasion. "It is nothing but the freedom to be wrong."

Besides, if the Earth was going around the Sun, nearby stars should seem to move against the background of more distant stars as, every six months, we shift our perspective from one side of the Earth's orbit to the other. No such "annual parallax" had been found. The Copernicans argued that this was because the stars were extremely far away—maybe a million times more distant than the Earth is from the Sun. Perhaps better telescopes, in future

---

* Copernicus' famous book was first published with an introduction by the theologian Andrew Osiander, inserted without the knowledge of the dying astronomer. Osiander's well-meaning attempt to reconcile religion and Copernican astronomy ended with these words: "[L]et no one expect anything in the way of certainty of astronomy, since astronomy can offer us nothing certain, lest, if anyone take as true that which has been constructed for another use, he go away from this discipline a bigger fool than when he came to it." Certainty could be found only in religion.

times, would find an annual parallax. The geocentrists considered this a desperate attempt to save a flawed hypothesis, and ludicrous on the face of it.

When Galileo turned the first astronomical telescope to the sky, the tide began to turn. He discovered that Jupiter had a little retinue of moons circling it, the inner ones orbiting faster than the outer ones, just as Copernicus had deduced for the motion of the planets about the Sun. He found that Mercury and Venus went through phases like the Moon (showing they orbited the Sun). Moreover, the cratered Moon and the spotted Sun challenged the perfection of the heavens. This may in part constitute the sort of trouble Tertullian was worried about thirteen hundred years earlier, when he pleaded, "If you have any sense or modesty, have done with prying into the regions of the sky, into the destiny and secrets of the universe."

In contrast, Galileo taught that we can interrogate Nature by observation and experiment. Then, "facts which at first sight seem improbable will, even on scant explanation, drop the cloak which had hidden them and stand forth in naked and simple beauty." Are not these facts, available even for skeptics to confirm, a surer insight into God's Universe than all the speculations of the theologians? But what if these facts contradict the beliefs of those who hold their religion incapable of making mistakes? The princes of the Church threatened the aged astronomer with torture if he persisted in teaching the abominable doctrine that the Earth moved. He was sentenced to a kind of house arrest for the remainder of his life.

A generation or two later, by the time Isaac Newton demonstrated that simple and elegant physics could quantitatively explain—and predict—all the observed lunar and planetary motions (provided you assumed the Sun at the center of the Solar System), the geocentrist conceit eroded further.

In 1725, in an attempt to discover stellar parallax, the painstaking English amateur astronomer James Bradley stumbled on the aberration of light. The term "aberration," I suppose, conveys something of the unexpectedness of the discovery. When observed over the course of a year, stars were found to trace little

ellipses against the sky. But all the stars were found to do so. This could not be stellar parallax, where we would expect a big parallax for nearby stars and an indetectible one for faraway stars. Instead, aberration is similar to how raindrops falling directly down on a speeding auto seem to the passengers to be falling at a slant; the faster the car goes, the steeper the slant. If the Earth were stationary at the center of the Universe, and not speeding in its orbit around the Sun, Bradley would not have found the aberration of light. It was a compelling demonstration that the Earth revolved about the Sun. It convinced most astronomers and some others but not, Bradley thought, the "Anti-Copernicans."

But not until 1837 did direct observations of the stars prove in the clearest way that the Earth is indeed circling the Sun. The long-debated annual parallax was at last discovered—not by better

The Horsehead Nebula and IC434. Courtesy ROE/Anglo-Australian Observatory. Photograph by David Malin.

arguments, but by better instruments. Because explaining what it means is much more straightforward than explaining the aberration of light, its discovery was very important. It pounded the final nail into the coffin of geocentrism. You need only look at your finger with your left eye and then with your right and see it seem to move. Everyone can understand parallax.

By the nineteenth century, all scientific geocentrists had been converted or rendered extinct. Once most scientists had been convinced, informed public opinion had swiftly changed, in some countries in a mere three or four generations. Of course, in the time of Galileo and Newton and even much later, there were still some who objected, who tried to prevent the new Sun-centered Universe from becoming accepted, or even known. And there were many who at least harbored secret reservations.

By the late twentieth century, just in case there were any holdouts, we have been able to settle the matter directly. We've been able to test whether we live in an Earth-centered system with planets affixed to transparent crystal spheres, or in a Sun-centered system with planets controlled at a distance by the gravity of the Sun. We have, for example, probed the planets with radar. When we bounce a signal off a moon of Saturn, we receive no radio echo from a nearer crystal sphere attached to Jupiter. Our spacecraft arrive at their appointed destinations with astonishing precision, exactly as predicted by Newtonian gravitation. When our ships fly to Mars, say, their instruments do not hear a tinkling sound or detect shards of broken crystal as they crash through the "spheres" that—according to the authoritative opinions that prevailed for millennia—propel Venus or the Sun in their dutiful motions about the central Earth.

When *Voyager 1* scanned the Solar System from beyond the outermost planet, it saw, just as Copernicus and Galileo had said we would, the Sun in the middle and the planets in concentric orbits about it. Far from being the center of the Universe, the Earth is just one of the orbiting dots. No longer confined to a single world, we are now able to reach out to others and determine decisively what kind of planetary system we inhabit.

EVERY OTHER PROPOSAL, and their number is legion, to displace us from cosmic center stage has also been resisted, in part for similar reasons. We seem to crave privilege, merited not by our works but by our birth, by the mere fact that, say, we are humans and born on Earth. We might call it the anthropocentric—the "human-centered"—conceit.

This conceit is brought close to culmination in the notion that we are created in God's image: The Creator and Ruler of the entire Universe looks just like me. My, what a coincidence! How convenient and satisfying! The sixth-century-B.C. Greek philosopher Xenophanes understood the arrogance of this perspective:

> The Ethiopians make their gods black and snub-nosed; the Thracians say theirs have blue eyes and red hair . . . Yes, and if oxen and horses or lions had hands, and could paint with their hands, and produce works of art as men do, horses would paint the forms of the gods like horses, and oxen like oxen . . .

Such attitudes were once described as "provincial"—the naive expectation that the political hierarchies and social conventions of an obscure province extend to a vast empire composed of many different traditions and cultures; that the familiar boondocks, *our* boondocks, are the center of the world. The country bumpkins know almost nothing about what else is possible. They fail to grasp the insignificance of their province or the diversity of the Empire. With ease, they apply their own standards and customs to the rest of the planet. But plopped down in Vienna, say, or Hamburg, or New York, ruefully they recognize how limited their perspective has been. They become "deprovincialized."

Modern science has been a voyage into the unknown, with a lesson in humility waiting at every stop. Many passengers would rather have stayed home.

THE GREAT DEMOTIONS

[One philosopher] asserted that he knew the whole secret . . . [H]e surveyed
the two celestial strangers from top to toe, and maintained to their faces that
their persons, their worlds, their suns, and their stars, were created solely for
the use of man. At this assertion our two travelers let themselves fall against
each other, seized with a fit of . . . inextinguishable laughter.

—VOLTAIRE, *MICROMEGAS. A PHILOSOPHICAL HISTORY* (1752)

OPPOSITE: The Universe of galaxies. This spectacular photograph obtained by the Hubble Space Telescope is of the outskirts of the Coma Cluster of galaxies, about 370 million light-years away. Virtually every object seen is a galaxy. Most prominent, at center right, is NGC 4881, a giant elliptical galaxy. The next largest, to its left, is a spiral galaxy like the Milky Way, seen face-on. The elongated galaxies are other spirals, seen roughly edge-on. The orange-and-white object trailing two tendrils is a pair of galaxies in collision; the gravity of each has distorted the form of the other. The black boxes represent an absence of data.

Many of the fainter galaxies in this image are not part of the Coma Cluster, but are sizable galaxies that appear faint because they are much farther away. Future generations of telescopes will capture the light of an enormously greater number of distant galaxies wholly unknown to us today.

The field of view of this picture is a little piece of the sky less than 1 percent of the apparent angular area of the Moon. It therefore represents only about one hundred millionth of the sky. The total number of stars in this field of view—the vast majority in other galaxies and too faint to be seen by Hubble—is in excess of one hundred trillion. The number of distant planets in this tiny piece of sky is, on the basis of modern evidence, comparably huge.

Each of these galaxies is spinning, typically making one rotation every few hundred million years. They are also in motion with respect to each other. The entire Coma Cluster, of which this is a tiny part, is moving with respect to other clusters of galaxies. Finally, all the galaxies in the Coma Cluster are collectively expanding away from all other clusters of galaxies. With respect to the Local Group of galaxies—of which the Milky Way is a part—the Coma Cluster is receding at about 7,000 kilometers per second. This is the motion called the expansion of the Universe that derives from the Big Bang.

Photo courtesy William A. Baum, the Hubble Telescope WFPC1 Team, and NASA .

In the seventeenth century there was still some hope that, even if the Earth was not the center of the Universe, it might be the only "world." But Galileo's telescope revealed that "the Moon certainly does not possess a smooth and polished surface" and that other worlds might look "just like the face of the Earth itself." The Moon and the planets showed unmistakably that they had as much claim to being worlds as the Earth does—with mountains, craters, atmospheres, polar ice caps, clouds, and, in the case of Saturn, a dazzling, unheard-of set of circumferential rings. After millennia of philosophical debate, the issue was settled decisively in favor of "the plurality of worlds." They might be profoundly different from our planet. None of them might be as congenial for life. But the Earth was hardly the only one.

This was the next in the series of Great Demotions, down-lifting experiences, demonstrations of our apparent insignificance, wounds that science has, in its search for Galileo's facts, delivered to human pride.

WELL, SOME HOPED, *even if the Earth isn't at the center of the Universe, the Sun is. The Sun is* our *Sun. So the Earth is* approximately *at the center of the Universe.* Perhaps some of our pride could in this way be salvaged. But by the nineteenth century, observational astronomy had made it clear that the Sun is but one lonely star in a great self-gravitating assemblage of suns called the Milky Way Galaxy. Far from being at the center of the Galaxy, our Sun with its entourage of dim and tiny planets lies in an undistinguished sector of an obscure spiral arm. We are thirty thousand light years from the Center.

Well, our Milky Way is the only galaxy. The Milky Way Galaxy is one of billions, perhaps hundreds of billions of galaxies notable neither in mass nor in brightness nor in how its stars are configured and arrayed. Some modern deep sky photographs show more galaxies beyond the Milky Way than stars within the Milky Way. Every one of them is an island universe containing perhaps a hundred billion suns. Such an image is a profound sermon on humility.

Well, then, at least our Galaxy is at the center of the Universe. No, this is wrong too. When the expansion of the Universe was first discovered, many people naturally gravitated to the notion that the

Milky Way was at the center of the expansion, and all the other galaxies running away from us. We now recognize that astronomers on any galaxy would see all the others running away from them; unless they were very careful, they would all conclude that *they* were at the center of the Universe. There is, in fact, *no* center to the expansion, no point of origin of the Big Bang, at least not in ordinary three-dimensional space.

Modern observations confirm Galileo's conclusion that the surface of the Moon is not smooth and polished. *Apollo 17* image, courtesy NASA.

Well, even if there are hundreds of billions of galaxies, each with hundreds of billions of stars, no other star has planets. If there are no other planets beyond our Solar System, perhaps there's no other life in the Universe. Our uniqueness might then be saved. Since planets are small and feebly shine by reflected sunlight, they're hard to find. Although applicable technology is improving with breathtaking speed, even a giant world like Jupiter, orbiting the *nearest* star, Alpha Centauri, would still be difficult to detect. In our ignorance, the geocentrists find hope.

There was once a scientific hypothesis—not just well received but prevailing—that supposed our solar system to have formed through the near collision of the ancient Sun with another star; the gravitational tidal interaction pulled out tendrils of sunstuff that quickly condensed into planets. Since space is mainly empty and near stellar collisions most rare, it was concluded that few other planetary systems exist—perhaps only one, around that other star that long ago co-parented the worlds of our solar system. Early in my studies, I was amazed and disappointed that such a view had ever been taken seriously, that for planets of other stars, absence of evidence had been considered evidence of absence.

Today we have firm evidence for at least three planets orbiting an extremely dense star, the pulsar designated B1257+12, about which I'll say more later. And we've found, for more than half the stars with masses like the Sun's, that early in their careers they're surrounded by great disks of gas and dust out of which planets seem to form. Other planetary systems now look to be a cosmic commonplace, maybe even worlds something like the Earth. We should be able, in the next few decades, to inventory at least the larger planets, if they exist, of hundreds of nearby stars.

Well, if our position in space doesn't reveal our special role, our posi-

tion in time does: We've been in the Universe since The Beginning (give or take a few days). *We've been given special responsibilities by the Creator.* It once seemed very reasonable to think of the Universe as beginning just a little before our collective memory is obscured by the passage of time and the illiteracy of our ancestors. Generally speaking, that's hundreds or thousands of years ago. Religions that purport to describe the origin of the Universe often specify—implicitly or explicitly—a date of origin of roughly such vintage, a birthday for the world.

If you add up all the "begats" in Genesis, for example, you get an age for the Earth: 6,000 years old, plus or minus a little. The universe is said to be exactly as old as the Earth. This is still the standard of Jewish, Christian, and Moslem fundamentalists and is clearly reflected in the Jewish calendar.

But so young a Universe raises an awkward question: How is it that there are astronomical objects more than 6,000 light-years away? It takes light a year to travel a light-year, 10,000 years to travel 10,000 light-years, and so on. When we look at the center of the Milky Way Galaxy, the light we see left its source 30,000 years ago. The nearest spiral galaxy like our own, M31 in the constellation Andromeda, is 2 million light-years away, so we are seeing it as it was when the light from it set out on its long journey to Earth—2 million years ago. And when we observe distant quasars 5 billion light-years away, we are seeing them as they were 5 billion years ago, before the Earth was formed. (They are, almost certainly, very different today.)

If, despite this, we were to accept the literal truth of such religious books, how could we reconcile the data? The only plausible conclusion, I think, is that God recently made all the photons of light arriving on the Earth in such a coherent format as to mislead generations of astronomers into the misapprehension that there are such things as galaxies and quasars, and intentionally driving them to the spurious conclusion that the Universe is vast and old.

The Milky Way Galaxy seen in infrared light from above the Earth's atmosphere. The spiral arms, of which our Sun is a part, are seen edge-on (because the Sun lies close to the plane of our galaxy). We are almost 30,000 light-years from the center. COBE image, courtesy NASA.

This is such a malevolent theology I still have difficulty believing that anyone, no matter how devoted to the divine inspiration of any religious book, could seriously entertain it.

Beyond this, the radioactive dating of rocks, the abundance of impact craters on many worlds, the evolution of the stars, and the expansion of the Universe each provides compelling and independent evidence that our Universe is many billions of years old—despite the confident assertions of revered theologians that a world so old directly contradicts the word of God, and that at any rate information on the antiquity of the world is inaccessible except to faith.★ These lines of evidence, as well, would have to be manufactured by a deceptive and malicious deity—unless the world is much older than the literalists in the Judeo-Christian-Islamic religion suppose. Of course, no such problem arises for those many religious people who treat the Bible and the Qur'an as historical and moral guides and great literature, but who recognize that the perspective of these scriptures on the natural world reflects the rudimentary science of the time in which they were written.

Ages rolled by before the Earth began. More ages will run their course before it is destroyed. A distinction needs to be drawn between how old the Earth is (around 4.5 billion years) and how old the Universe is (about 15 billion years since the Big Bang). The immense interval of time between the origin of the Universe and our epoch was two-thirds over before the Earth came to be. Some stars and planetary systems are billions of years younger, others billions of years older. But in Genesis, chapter 1, verse 1, the Universe and the Earth are created on the same day. The Hindu-Buddhist-Jain religion tends not to confound the two events.

★ St. Augustine, in *The City of God,* says, "As it is not yet six thousand years since the first man . . . are not those to be ridiculed rather than refuted who try to persuade us of anything regarding a space of time so different from, and contrary to, the ascertained truth? . . . We, being sustained by divine authority in the history of our religion, have no doubt that whatever is opposed to it is most false." He excoriates the ancient Egyptian tradition that the world is as much as a hundred thousand years old as "abominable lies." St. Thomas Aquinas, in the *Summa Theologica,* flatly states that "the newness of the world cannot be demonstrated from the world itself." They were so *sure.*

As for humans, we're latecomers. We appear in the last instant of cosmic time. The history of the Universe till now was 99.998 percent over before our species arrived on the scene. In that vast sweep of aeons, we could not have assumed any special responsibilities for our planet, or life, or anything else. We were not here.

Well, if we can't find anything special about our position or our epoch, maybe there's something special about our motion. Newton and all the other great classical physicists held that the velocity of the Earth in space constituted a "privileged frame of reference." That's actually what it was called. Albert Einstein, a keen critic of prejudice and privilege all his life, considered this "absolute" physics a remnant of an increasingly discredited Earth chauvinism. It seemed to him that the laws of Nature must be the same no matter what the velocity or frame of reference of the observer. With this as his starting point, he developed the Special Theory of Relativity. Its consequences are bizarre, counterintuitive, and grossly contradict common sense—but only at very high speeds. Careful and repeated observations show that his justly celebrated theory is an accurate description of how the world is made. Our common-sense intuitions can be mistaken. Our preferences don't count. We do not live in a privileged reference frame.

One consequence of special relativity is time dilation—the slowing down of time as the observer approaches light speed. You can still find claims that time dilation applies to watches and elementary particles—and, presumably, to circadian and other rhythms in plants, animals, and microbes—but not to human biological clocks. Our species has been granted, it is suggested, special immunity from the laws of Nature, which must accordingly be able to distinguish deserving from undeserving collections of matter. (In fact, the proof Einstein gave for special relativity admits no such distinctions.) The idea of humans as exceptions to relativity seems another incarnation of the notion of special creation:

Well, even if our position, our epoch, our motion, and our world are not unique, maybe we are. We're different from the other animals. We're specially created. The particular devotion of the Creator of the Universe is evident in us. This position was passionately defended on religious and other grounds. But in the middle nineteenth century Charles Darwin showed convincingly how one species can evolve into another by entirely natural processes, which come down to the heartless business of Nature saving the heredities that work and rejecting those that don't. "Man in his arrogance thinks himself a great work worthy [of] the interposition of a deity," Darwin wrote telegraphically in his notebook. "More humble and I think truer to consider him created from animals." The profound and intimate connections of humans with the other life forms on Earth have been compellingly demonstrated in the late twentieth century by the new science of molecular biology.

IN EACH AGE the self-congratulatory chauvinisms are challenged in yet another arena of scientific debate—in this century, for example, in attempts to understand the nature of human sexuality, the existence of the unconscious mind, and the fact that many psychiatric illnesses and character "defects" have a molecular origin. But also:

Well, even if we're closely related to some of the other animals, we're different—not just in degree, but in kind—on what really matters: reasoning, self-consciousness, tool making, ethics, altruism, religion, language, nobility of character. While humans, like all animals, have traits that set them apart—otherwise, how could we distinguish one species from another?—human uniqueness has been exaggerated, sometimes grossly so. Chimps reason, are self-conscious, make tools, show devotion, and so on. Chimps and humans have 99.6 percent of their active genes in common. (Ann Druyan and I run through the evidence in our book *Shadows of Forgotten Ancestors*.)

In popular culture, the very opposite position is also embraced, although it too is driven by human chauvinism (plus a failure of the imagination): Children's stories and cartoons make animals dress in clothes, live in houses, use knives and forks, and speak. The three bears sleep in beds. The owl and the pussycat go to sea in a beautiful pea-green boat. Dinosaur mothers cuddle their young. Pelicans deliver the mail. Dogs drive cars. A worm catches a thief. Pets have human names. Dolls, nutcrackers, cups, and saucers dance and have opinions. The dish runs away with the spoon. In the *Thomas the Tank Engine* series, we even have anthropomorphic locomotives and railway cars, charmingly portrayed. No matter what we're thinking about, animate or inanimate, we tend to invest it with human traits. We can't help ourselves. The images come readily to mind. Children are clearly fond of them.

When we talk about a "threatening" sky, a "troubled" sea, diamonds "resisting" being scratched, the Earth "attracting" a passing asteroid, or an atom being "excited," we are again drawn to a kind of animist worldview. We reify. Some ancient level of our thinking endows inanimate Nature with life, passions, and forethought.

The notion that the Earth is self-aware has lately been growing at the fringes of the "Gaia" hypothesis. But this was a commonplace belief of both the ancient Greeks and the early Christians. Origen wondered whether "the earth also, according to its own nature, is accountable for some sin." A host of ancient scholars thought the stars alive. This was also the position of Origen, of St. Ambrose (the mentor of St. Augustine), and even, in a more qualified form, of St. Thomas Aquinas. The Stoic philosophical position on the Sun's nature was stated by Cicero, in the first century B.C.: "Since the Sun resembles those fires which are contained in the bodies of living creatures, the Sun must also be alive."

Animist attitudes in general seem to have been spreading recently. In a 1954 American survey, 75 percent of people polled were willing to state that the Sun is not alive; in 1989, only 30 percent would support so rash a proposition. On whether an automobile tire can feel anything, 90 percent of respondents denied it emotions in 1954, but only 73 percent in 1989.

We can recognize here a shortcoming—in some circumstances serious—in our ability to understand the world. Characteristically, willy-nilly, we seem compelled to project our own nature onto Nature. Although this may result in a consistently distorted view of the world, it does have one great virtue—projection is the essential precondition for compassion.

Okay, maybe we're not much, maybe we're humiliatingly related to apes, but at least we're the best there is. God and angels aside, we're the only intelligent beings in the Universe. One correspondent writes to me, "I am as sure of this as anything in my experience. There is no conscious life anywhere else in the Universe. Mankind thus returns to its rightful position as center of the Universe." However, partly through the influence of science and science fiction, most people today, in the United States at least, reject this proposition—

for reasons essentially stated by the ancient Greek philosopher Chrysippus: "For any human being in existence to think that there is nothing in the whole world superior to himself would be an insane piece of arrogance."

But the simple fact is that we have not yet found extraterrestrial life. We are in the earliest stages of looking. The question is wide open. If I had to guess—especially considering our long sequence of failed chauvinisms—I would guess that the Universe is filled with beings far more intelligent, far more advanced than we are. But of course I might be wrong. Such a conclusion is at best based on a plausibility argument, derived from the numbers of planets, the ubiquity of organic matter, the immense timescales available for evolution, and so on. It is not a scientific demonstration. The question is among the most fascinating in all of science. As described in this book, we are just developing the tools to treat it seriously.

What about the related matter of whether we are capable of creating intelligences smarter than ourselves? Computers routinely do mathematics that no unaided human can manage, outperform world champions in checkers and grand masters in chess, speak and understand English and other languages, write presentable short stories and musical compositions, learn from their mistakes, and competently pilot ships, airplanes, and spacecraft. Their abilities steadily improve. They're getting smaller, faster, and cheaper. Each year, the tide of scientific advance laps a little further ashore on the island of human intellectual uniqueness with its embattled castaways. If, at so early a stage in our technological evolution, we have been able to go so far in creating intelligence out of silicon and metal, what will be possible in the following decades and centuries? What happens when smart machines are able to manufacture smarter machines?

PERHAPS THE CLEAREST INDICATION that the search for an unmerited privileged position for humans will never be wholly abandoned is what in physics and astronomy is called the Anthropic Principle. It would be better named the Anthropocentric Principle. It comes in various forms. The "Weak" Anthropic Principle merely notes that if the laws of Nature and the physical constants—such as the speed

of light, the electrical charge of the electron, the Newtonian gravitational constant, or Planck's quantum mechanical constant—had
been different, the course of events leading to the origin of
humans would never have transpired. Under other laws and constants, atoms would not hold together, stars would evolve too
quickly to leave sufficient time for life to evolve on nearby planets, the chemical elements of which life is made would never have
been generated, and so on. Different laws, no humans.

There is no controversy about the Weak Anthropic Principle:
Change the laws and constants of Nature, if you could, and a very
different universe may emerge—in many cases, a universe incompatible with life.[*] The mere fact that we exist implies (but does
not impose) constraints on the laws of Nature. By contrast, the
various "Strong" Anthropic Principles go much further; some of
their advocates come close to deducing that the laws of Nature
and the values of the physical constants were established (don't ask
how or by Whom) so *that* humans would eventually come to be.
Almost all of the other possible universes, they say, are inhos-
pitable. In this way, the ancient conceit that the Universe was
made for us is resuscitated.

To me it echoes Dr. Pangloss in Voltaire's *Candide,* convinced
that this world, with all its imperfections, is the best possible. It
sounds like playing my first hand of bridge, winning, knowing that
there are 54 billion billion billion (5.4×10^{28}) possible other hands
that I was equally likely to have been dealt . . . and then foolishly
concluding that a god of bridge exists and favors me, a god who
arranged the cards and the shuffle with my victory foreordained
from The Beginning. We do not know how many other winning
hands there are in the cosmic deck, how many other kinds of universes, laws of Nature, and physical constants that could also lead

---

[*] Our universe is *almost* incompatible with life—or at least what we understand
as necessary for life: Even if every star in a hundred billion galaxies had an
Earthlike planet, without heroic technological measures life could prosper in
only about 10^{-37} the volume of the Universe. For clarity, let's write it out: only
0.000 000 000 000 000 000 000 000 000 000 000 000 1 of our universe is hospitable to life. Thirty-six zeroes before the one. The rest is cold, radiation-rid-
dled black vacuum.

to life and intelligence and perhaps even delusions of self-importance. Since we know next to nothing about how the Universe
was made—or even *if* it was made—it's difficult to pursue these
notions productively.

Voltaire asked "Why is there anything?" Einstein's formula-
tion was to ask whether God had any choice in creating the Universe. But if the Universe is infinitely old—if the Big Bang some
15 billion years ago is only the most recent cusp in an infinite series of cosmic contractions and expansions—then it was never created and the question of why it is as it is is rendered meaningless.

If, on the other hand, the Universe has a finite age, why is it
the way it is? Why wasn't it given a very different character?
Which laws of Nature go with which others? Are there meta-laws
specifying the connections? Can we possibly discover them? Of all
conceivable laws of gravity, say, which ones can exist simultaneously with which conceivable laws of quantum physics that determine the very existence of macroscopic matter? Are all laws we
can think of possible, or is there only a restricted number that can
somehow be brought into existence? Clearly we have not a glim-
mering of how to determine which laws of Nature are "possible"
and which are not. Nor do we have more than the most rudimentary notion of what correlations of natural laws are "permitted."

For example, Newton's universal law of gravitation specifies
that the mutual gravitational force attracting two bodies towards
each other is inversely proportional to the square of how far they
are apart. You move twice as far from the center of the Earth and
you weigh a quarter as much; ten times farther and you weigh
only a hundredth of your ordinary weight; etc. It is this inverse
square law that permits the exquisite circular and elliptical orbits
of planets around the Sun, and moons around the planets—as well
as the precision trajectories of our interplanetary spacecraft. If r is
the distance between the centers of two masses, we say that the
gravitational force varies as $1/r^2$.

But if this exponent were different—if the gravitational law
were $1/r^4$, say, rather than $1/r^2$—then the orbits would not close;
over billions of revolutions, the planets would spiral in and be con-
sumed in the fiery depths of the Sun, or spiral out and be lost to

interstellar space. If the Universe were constructed with an inverse fourth power law rather than an inverse square law, soon there would be no planets for living beings to inhabit.

So of all the possible gravitational force laws, why are we so lucky as to live in a universe sporting a law consistent with life? First, of course, we're so "lucky," because if we weren't, we wouldn't be here to ask the question. It is no mystery that inquisitive beings who evolve on planets can be found only in universes that admit planets. Second, the inverse square law is *not* the only one consistent with stability over billions of years. Any power law less steep than $1/r^3$ ($1/r^{2.99}$ or $1/r$, for example) will keep a planet in the *vicinity* of a circular orbit even if it's given a shove. We have a tendency to overlook the possibility that other conceivable laws of Nature might also be consistent with life.

But there's a further point: It's not arbitrary that we have an inverse square law of gravitation. When Newton's theory is understood in terms of the more encompassing general theory of relativity, we recognize that the exponent of the gravity law is 2 because the number of physical dimensions we live in is 3. All gravity laws aren't available, free for a Creator's choosing. Even given an infinite number of three-dimensional universes for some great god to tinker with, the gravity law would always have to be the law of the inverse square. Newtonian gravity, we might say, is not a contingent facet of our universe, but a necessary one.

In general relativity, gravity is *due to* the dimensionality and curvature of space. When we talk about gravity we are talking about local dimples in space-time. This is by no means obvious and even affronts commonsense notions. But when examined deeply, the ideas of gravity and mass are not separate matters, but ramifications of the underlying geometry of space-time.

I wonder if something like this doesn't apply generally to all anthropic hypotheses. The laws or physical constants on which our lives depend turn out to be members of a class, perhaps even a vast class, of other laws and other physical constants—but some of these are also compatible with a kind of life. Often we do not (or cannot) work through what those other universes allow. Beyond that, not every arbitrary choice of a law of Nature or a physical constant may be available, even to a maker of universes. Our un-

The barred spiral galaxy
NGC 1365. Courtesy
Anglo-Australian Observatory.
Photograph by David Malin.

derstanding of which laws of Nature and which physical constants
are up for grabs is fragmentary at best.

Moreover, we have no access to any of those putative alternative universes. We have no experimental method by which anthropic hypotheses may be tested. Even if the existence of such
universes were to follow firmly from well-established theories—of
quantum mechanics or gravitation, say—we could not be sure that
there weren't better theories that predict no alternative universes.
Until that time comes, if it ever does, it seems to me premature to
put faith in the Anthropic Principle as an argument for human
centrality or uniqueness.

Finally, even if the Universe were *intentionally* created to
allow for the emergence of life or intelligence, other beings may
exist on countless worlds. If so, it would be cold comfort to an-

thropocentrists that we inhabit one of the few universes that allow life and intelligence.

There is something stunningly narrow about how the Anthropic Principle is phrased. Yes, only certain laws and constants of nature are consistent with our kind of life. But essentially the same laws and constants are required to make a rock. So why not talk about a Universe designed so rocks could one day come to be, and strong and weak Lithic Principles? If stones could philosophize, I imagine Lithic Principles would be at the intellectual frontiers.

There are cosmological models being formulated today in which even the entire Universe is nothing special. Andrei Linde, formerly of the Lebedev Physical Institute in Moscow and now at Stanford University, has incorporated current understanding of the strong and weak nuclear forces and quantum physics into a new cosmological model. Linde envisions a vast Cosmos, much larger than our Universe—perhaps extending to infinity both in space and time—not the paltry 15 billion light-years or so in radius and 15 billion years in age which are the usual understanding. In this Cosmos there is, as here, a kind of quantum fluff in which tiny structures much smaller than an electron are everywhere forming, reshaping, and dissipating; in which, as here, fluctuations in absolutely empty space create pairs of elementary particles—an electron and a positron, for example. In the froth of quantum bubbles, the vast majority remain submicroscopic. But a tiny fraction inflate, grow, and achieve respectable universehood. They are so far away from us, though—much farther than the 15 billion light-years that is the conventional scale of our universe—that, if they exist, they appear to be wholly inaccessible and undetectable.

Most of these other universes reach a maximum size and then collapse, contract to a point, and disappear forever. Others may oscillate. Still others may expand without limit. In different universes there will be different laws of nature. We live, Linde argues, in one such universe—one in which the physics is congenial for growth, inflation, expansion, galaxies, stars, worlds, life. We imagine our universe to be unique, but it is one of an immense number—perhaps an infinite number—of equally valid, equally independent, equally isolated universes. There will be life in some, and not in others. In this view the observable Universe is just a newly formed backwater of a much vaster, infinitely old, and wholly unobservable Cosmos. If something like this is right, even our residual pride, pallid as it must be, of living in the only universe is denied to us.*

Maybe someday, despite current evidence, a means will be devised to peer into adjacent universes, sporting very different laws of nature, and we will see what else is possible. Or perhaps inhabitants of adjacent universes can peer into ours. Of course, in such speculations we have far exceeded the bounds of knowledge. But if something like Linde's Cosmos is true, there is—amazingly—still another devastating deprovincialization awaiting us.

Our powers are far from adequate to be creating universes anytime soon. Strong Anthropic Principle ideas are not amenable to proof (although Linde's cosmology does have some testable features). Extraterrestrial life aside, if self-congratulatory pretensions to centrality have now retreated to such bastions impervious to experiment, then the sequence of scientific battles with human chauvinism would seem to have been, at least largely, won.

THE LONG-STANDING VIEW, as summarized by the philosopher Immanuel Kant, that "without man . . . the whole of creation would be a mere wilderness, a thing in vain, and have no final end" is revealed to be self-indulgent folly. A Principle of Mediocrity seems to apply to all our circumstances. We could not have known beforehand that the evidence would be, so repeatedly and thoroughly, incompatible with the proposition that human beings are at center stage in the Universe. But most of the debates have now been settled decisively in favor of a position that, however painful, can be encapsulated in a single sentence: We have not been given the lead in the cosmic drama.

Perhaps someone else has. Perhaps no one else has. In either case, we have good reason for humility.

---

* For such ideas, words tend to fail us. A German locution for Universe is [das] All—which makes the inclusiveness quite unmistakable. We might say that our universe is but one in a "Multiverse," but I prefer to use "Cosmos" for everything and "Universe" for the only one we can know about.

A UNIVERSE NOT MADE FOR US

The Sea of Faith
Was once, too, at the full, and round earth's shore
Lay like the folds of a bright girdle furl'd.
But now I only hear
Its melancholy, long, withdrawing roar,
Retreating, to the breath
Of the night-wind, down the vast edges drear
And naked shingles of the world.
—MATTHEW ARNOLD, "DOVER BEACH" (1867)

"What a beautiful sunset," we say, or "I'm up before sunrise." No matter what the scientists allege, in everyday speech we often ignore their findings. We don't talk about the Earth turning, but about the Sun rising and setting. Try formulating it in Copernican language. Would you say, "Billy, be home by the time the Earth has rotated enough so as to occult the Sun below the local horizon"? Billy would be long gone before you're finished. We haven't been able even to find a graceful locution that accurately

OPPOSITE: "The Fabric of Space," watercolor by Greg Mort. The Universe as an unexpected gift.

conveys the heliocentric insight. We at the center and everything else circling us is built into our languages; we teach it to our children. We are unreconstructed geocentrists hiding behind a Copernican veneer.*

In 1633 the Roman Catholic Church condemned Galileo for teaching that the Earth goes around the Sun. Let's take a closer look at this famous controversy. In the preface to his book comparing the two hypotheses—an Earth-centered and a Sun-centered universe—Galileo had written,

> The celestial phenomena will be examined, strengthening the Copernican hypothesis until it might seem that this must triumph absolutely.

And later in the book he confessed,

> Nor can I ever sufficiently admire [Copernicus and his followers]; they have through sheer force of intellect done such violence to their own senses as to prefer what reason told them over what sensible experience plainly showed them . . .

The Church declared, in its indictment of Galileo,

> The doctrine that the earth is neither the center of the universe nor immovable, but moves even with a daily rotation, is absurd, and both psychologically and theologically false, and at the least an error of faith.

Galileo replied,

> The doctrine of the movements of the earth and the fixity of the sun is condemned on the ground that the Scriptures speak in many places of the sun moving and the earth standing still . . .

* One of the few quasi-Copernican expressions in English is "The Universe doesn't revolve around *you*"—an astronomical truth intended to bring fledgling narcissists down to Earth.

It is piously spoken that the Scriptures cannot lie. But none will deny that they are frequently abstruse and their true meaning difficult to discover, and more than the bare words signify. I think that in the discussion of natural problems we ought to begin not with the Scriptures, but with experiments and demonstrations.

But in his recantation (June 22, 1633) Galileo was made to say,

Having been admonished by the Holy Office entirely to abandon the false opinion that the Sun was the center of the universe and immovable, and that the Earth was not the center of the same and that it moved . . . I have been . . . suspected of heresy, that is, of having held and believed that the Sun is the center of the universe and immovable, and that the Earth is not the center of the same, and that it does move . . . I abjure with a sincere heart and unfeigned faith, I curse and detest the same errors and heresies, and generally all and every error and sect contrary to the Holy Catholic Church.

It took the Church until 1832 to remove Galileo's work from its list of books which Catholics were forbidden to read at the risk of dire punishment of their immortal souls.

Pontifical disquiet with modern science has ebbed and flowed since the time of Galileo. The high-water mark in recent history is the 1864 *Syllabus of Errors* of Pius IX, the pope who also convened the Vatican Council at which the doctrine of papal infallibility was, at his insistence, first proclaimed. Here are a few excerpts:

Divine revelation is perfect and, therefore, it is not subject to continual and indefinite progress in order to correspond with the progress of human reason . . . No man is free to embrace and profess that religion which he believes to be true, guided by the light of reason . . . The Church has power to define dogmatically the religion of the Catholic Church to be the only true religion . . . It is necessary even in the present day that the

Catholic religion shall be held as the only religion of the state, to the exclusion of all other forms of worship . . . The civil liberty of every mode of worship, and full power given to all of openly and publicly manifesting their opinions and their ideas conduce more easily to corrupt the morals and minds of the people . . . The Roman Pontiff cannot and ought not to reconcile himself or agree with, progress, liberalism and modern civilization.

To its credit, although belatedly and reluctantly, the Church in 1992 repudiated its denunciation of Galileo. It still cannot quite bring itself, though, to see the significance of its opposition. In a 1992 speech Pope John Paul II argued,

From the beginning of the Age of Enlightenment down to our own day, the Galileo case has been a sort of "myth" in which the image fabricated out of the events is quite far removed from reality. In this perspective, the Galileo case was a symbol of the Catholic Church's supposed rejection of scientific progress, or of "dogmatic" obscurantism opposed to the free search for truth.

But surely the Holy Inquisition ushering the elderly and infirm Galileo in to inspect the instruments of torture in the dungeons of the Church not only admits but requires just such an interpretation. This was not mere scientific caution and restraint, a reluctance to shift a paradigm until compelling evidence, such as the annual parallax, was available. This was fear of discussion and debate. Censoring alternative views and threatening to torture their proponents betray a lack of faith in the very doctrine and parishioners that are ostensibly being protected. Why were threats and Galileo's house arrest needed? Cannot truth defend itself in its confrontation with error?

The Pope does, though, go on to add:

The error of the theologians of the time, when they maintained the centrality of the earth, was to think that our understanding of the physical world's structure was in some way imposed by the literal sense of Sacred Scriptures.

OPPOSITE: A change of perspective: The Earth rising above the ancient impact basin Mare Smythii on the Moon. If we lived on the Moon, would we consider it the center of the Universe—with the Earth paying homage to us? *Apollo 11* photographs, courtesy NASA.

Here indeed considerable progress has been made—although proponents of fundamentalist faiths will be distressed to hear from the Pontiff that Sacred Scripture is not always literally true.

But if the Bible is not everywhere literally true, which parts are divinely inspired and which are merely fallible and human? As soon as we admit that there are scriptural mistakes (or concessions to the ignorance of the times), then how can the Bible be an inerrant guide to ethics and morals? Might sects and individuals now accept as authentic the parts of the Bible they like, and reject those that are inconvenient or burdensome? Prohibitions against murder, say, are essential for a society to function, but if divine retribution for murder is considered implausible, won't more people think they can get away with it?

Many felt that Copernicus and Galileo were up to no good and erosive of the social order. Indeed any challenge, from any source, to the literal truth of the Bible might have such consequences. We can readily see how science began to make people nervous. Instead of criticizing those who perpetuated the myths, public rancor was directed at those who discredited them.

OUR ANCESTORS UNDERSTOOD origins by extrapolating from their own experience. How else could they have done it? So the Universe was hatched from a cosmic egg, or conceived in the sexual congress of a mother god and a father god, or was a kind of product of the Creator's workshop—perhaps the latest of many flawed attempts. And the Universe was not much bigger than we see, and not much older than our written or oral records, and nowhere very different from places that we know.

We've tended in our cosmologies to make things familiar. Despite all our best efforts, we've not been very inventive. In the West, Heaven is placid and fluffy, and Hell is like the inside of a volcano. In many stories, both realms are governed by dominance hierarchies headed by gods or devils. Monotheists talked about the king of kings. In every culture we imagined something like our own political system running the Universe. Few found the similarity suspicious.

Then science came along and taught us that we are not the

measure of all things, that there are wonders unimagined, that the Universe is not obliged to conform to what we consider comfortable or plausible. We have learned something about the idiosyncratic nature of our common sense. Science has carried human self-consciousness to a higher level. This is surely a rite of passage, a step towards maturity. It contrasts starkly with the childishness and narcissism of our pre-Copernican notions.

But why should we *want* to think that the Universe was made for us? Why is the idea so appealing? Why do we nurture it? Is our self-esteem so precarious that nothing short of a universe custom-made for us will do?

Of course it appeals to our vanity. "What a man desires, he also imagines to be true," said Demosthenes. "The light of faith makes us see what we believe," cheerfully admitted St. Thomas Aquinas. But I think there may be something else. There's a kind of ethnocentrism among primates. To whichever little group we happen to be born, we owe passionate love and loyalty. Members of other groups are beneath contempt, deserving of rejection and hostility. That both groups are of the same species, that to an outside observer they are virtually indistinguishable, makes no difference. This is certainly the pattern among the chimpanzees, our closest relatives in the animal kingdom. Ann Druyan and I have described how this way of viewing the world may have made enormous evolutionary sense a few million years ago, however dangerous it has become today. Even members of hunter-gatherer groups—as far from the technological feats of our present global civilization as it is possible for humans to be—solemnly describe their little band, whichever it is, as "the people." Everyone else is something different, something less than human.

If this is our natural way of viewing the world, then it should occasion no surprise that every time we make a naive judgment about our place in the Universe—one untempered by careful and skeptical scientific examination—we almost always opt for the centrality of our group and circumstance. We want to believe, moreover, that these are objective facts, and not our prejudices finding a sanctioned vent.

So it's not much fun to have a gaggle of scientists incessantly haranguing us with "You're ordinary, you're unimportant, your privileges are undeserved, there's nothing special about you." Even unexcitable people might, after a while, grow annoyed at this incantation and those who insist on chanting it. It almost seems that the scientists are getting some weird satisfaction out of putting humans down. Why can't they find a way in which we're superior? Lift our spirits! Exalt us! In such debates science, with its mantra of discouragement, feels cold and remote, dispassionate, detached, unresponsive to human needs.

And, again, if we're not important, not central, not the apple of God's eye, what is implied for our theologically based moral codes? The discovery of our true bearings in the Cosmos was resisted for so long and to such a degree that many traces of the debate remain, sometimes with the motives of the geocentrists laid bare. Here, for example, is a revealing unsigned commentary in the British review *The Spectator* in 1892:

> [I]t is certain enough that the discovery of the heliocentric motion of the planets which reduced our earth to its proper "insignificance" in the solar system, did a good deal to reduce to a similar but far from proper "insignificance" the moral principles by which the predominant races of the earth had hitherto been guided and restrained. Part of this effect was no doubt due to the evidence afforded that the physical science of various inspired writers was erroneous instead of being infallible,—a conviction which unduly shook the confidence felt even in their moral and religious teaching. But a good deal of it was due only to the mere sense of "insignificance" with which man has contemplated himself, since he has discovered that he inhabits nothing but a very obscure corner of the universe, instead of the central world round which sun, moon, and stars alike revolved. There can be no doubt that man may feel himself, and has often felt himself, a great deal too insignificant to be the object of any particular divine training or care. If the earth be regarded as a sort of ant-hill, and the life and death of human beings as the life and death of so many ants which run into and out of so many holes in search of food and sunlight, it is quite certain that no ade-

quate importance will be attached to the duties of human life, and that a profound fatalism and hopelessness, instead of new hopefulness, will attach to human effort . . .

[F]or the present at least, our horizons are quite vast enough . . . ; till we can get used to the infinite horizons we already have, and not lose our balance so much as we usually do in contemplating them, the yearning for still wider horizons is premature.

WHAT DO WE REALLY WANT from philosophy and religion? Palliatives? Therapy? Comfort? Do we want reassuring fables or an understanding of our actual circumstances? Dismay that the Universe does not conform to our preferences seems childish. You might think that grown-ups would be ashamed to put such disappointments into print. The fashionable way of doing this is not to blame the Universe—which seems truly pointless—but rather to blame the means by which we know the Universe, namely science.

George Bernard Shaw, in the preface to his play *St. Joan,* described a sense of science preying on our credulity, forcing on us an alien worldview, intimidating belief:

> In the Middle Ages, people believed that the Earth was flat, for which they had at least the evidence of their senses: we believe it to be round, not because as many as one per cent of us could give the physical reason for so quaint a belief, but because modern science has convinced us that nothing that is obvious is true, and that everything that is magical, improbable, extraordinary, gigantic, microscopic, heartless, or outrageous is scientific.

A more recent and very instructive example is *Understanding the Present: Science and the Soul of Modern Man,* by Bryan Appleyard, a British journalist. This book makes explicit what many people feel, all over the world, but are too embarrassed to say. Appleyard's candor is refreshing. He is a true believer and will not let us slough over the contradictions between modern science and traditional religion:

"Science has taken away our religion," he laments. And what

sort of religion is it that he longs for? One in which "the human race was the point, the heart, the final cause of the whole system. It placed our selves definitively upon the universal map." . . . "We were the end, the purpose, the rational axle around which the great aetherian shells rotated." He longs for "the universe of Catholic orthodoxy" in which "the cosmos is shown to be a machine constructed around the drama of salvation"—by which Appleyard means that, despite explicit orders to the contrary, a woman and a man once ate of an apple, and that this act of insubordination transformed the Universe into a contrivance for operant-conditioning their remote descendants.

By contrast, modern science "presents us as accidents. We are caused by the cosmos, but we are not the cause of it. Modern man is not finally anything, he has no role in creation." Science is "spiritually corrosive, burning away ancient authorities and traditions. It cannot really co-exist with anything." . . . "Science, quietly and inexplicitly, is talking us into abandoning our selves, our true selves." It reveals "the mute, alien spectacle of nature." . . . "Human beings cannot live with such a revelation. The only morality left is that of the consoling lie." Anything is better than grappling with the unbearable burden of being tiny.

In a passage reminiscent of Pius IX, Appleyard even decries the fact that "a modern democracy can be expected to include a number of contradictory religious faiths which are obliged to agree on a certain limited number of general injunctions, but no more. They must not burn each other's places of worship, but they may deny, even abuse each other's God. This is the effective, scientific way of proceeding."

But what is the alternative? Obdurately to pretend to certainty in an uncertain world? To adopt a comforting belief system, no matter how out of kilter with the facts it is? If we don't know what's real, how can we deal with reality? For practical reasons, we cannot live too much in fantasyland. Shall we censor one another's religions and burn down one another's places of worship? How can we be sure which of the thousands of human belief systems should become unchallenged, ubiquitous, mandatory?

These quotations betray a failure of nerve before the Uni-

verse—its grandeur and magnificence, but especially its indiffer-
ence. Science has taught us that, because we have a talent for de-
ceiving ourselves, subjectivity may not freely reign. This is one
reason Appleyard so mistrusts science: It seems too reasoned, measured, and impersonal. Its conclusions derive from the interroga-
tion of Nature, and are not in all cases predesigned to satisfy our
wants. Appleyard deplores moderation. He yearns for inerrant doctrine, release from the exercise of judgment, and an obligation to
believe but not to question. He has not grasped human fallibility.
He recognizes no need to institutionalize error-correcting machinery either in our social institutions or in our view of the Universe.

   This is the anguished cry of the infant when the Parent does
not come. But most people eventually come to grips with reality,
and with the painful absence of parents who will absolutely guarantee that no harm befalls the little ones so long as they do what
they are told. Eventually most people find ways to accommodate
to the Universe—especially when given the tools to think straight.

   "All that we pass on to our children" in the scientific age,
Appleyard complains, "is the conviction that nothing is true, final
or enduring, including the culture from which they sprang." How
right he is about the inadequacy of our legacy. But would it be en-
riched by adding baseless certainties? He scorns "the pious hope
that science and religion are independent realms which can easily
be separated." Instead, "science, as it is now, is absolutely not compatible with religion."

   But isn't Appleyard really saying that some religions now find
it difficult to make unchallenged pronouncements on the nature
of the world that are straight-out false? We recognize that even
revered religious leaders, the products of their time as we are of
ours, may have made mistakes. Religions contradict one another—on small matters, such as whether we should put on a hat
or take one off on entering a house of worship, or whether we
should eat beef and eschew pork or the other way around, all the
way to the most central issues, such as whether there are no gods,
one God, or many gods.

   Science has brought many of us to that state in which

Nathaniel Hawthorne found Herman Melville: "He can neither believe, nor be comfortable in his unbelief." Or Jean-Jacques Rousseau: "They had not persuaded me, but they had troubled me. Their arguments had shaken me without ever convincing me . . . It is hard to prevent oneself from believing what one so keenly desires." As the belief systems taught by the secular and religious authorities are undermined, respect for authority in general probably does erode. The lesson is clear: Even political leaders must be wary of embracing false doctrine. This is not a failing of science, but one of its graces.

Of course, worldview consensus is comforting, while clashes of opinion may be unsettling, and demand more of us. But unless we insist, against all evidence, that our ancestors were perfect, the advance of knowledge requires us to unravel and then restitch the consensus they established.

In some respects, science has far surpassed religion in delivering awe. How is it that hardly any major religion has looked at science and concluded, "This is better than we thought! The Universe is much bigger than our prophets said, grander, more subtle, more elegant. God must be even greater than we dreamed"? Instead they say, "No, no, no! My god is a little god, and I want him to stay that way." A religion, old or new, that stressed the magnificence of the Universe as revealed by modern science might be able to draw forth reserves of reverence and awe hardly tapped by the conventional faiths. Sooner or later, such a religion will emerge.

IF YOU LIVED two or three millennia ago, there was no shame in holding that the Universe was made for us. It was an appealing thesis consistent with everything we knew; it was what the most learned among us taught without qualification. But we have found out much since then. Defending such a position today amounts to willful disregard of the evidence, and a flight from self-knowledge.

Still, for many of us, these deprovincializations rankle. Even if they do not fully carry the day, they erode confidence—unlike the happy anthropocentric certitudes, rippling with social utility, of an earlier age. We long to be here for a purpose, even though, despite much self-deception, none is evident. "The meaningless absurdity of life," wrote Leo Tolstoy, "is the only incontestable knowledge accessible to man." Our time is burdened under the cumulative weight of successive debunkings of our conceits: We're Johnny-come-latelies. We live in the cosmic boondocks. We emerged from microbes and muck. Apes are our cousins. Our thoughts and feelings are not fully under our own control. There may be much smarter and very different beings elsewhere. And on top of all this, we're making a mess of our planet and becoming a danger to ourselves.

The trapdoor beneath our feet swings open. We find ourselves in bottomless free fall. We are lost in a great darkness, and there's no one to send out a search party. Given so harsh a reality, of course we're tempted to shut our eyes and pretend that we're safe and snug at home, that the fall is only a bad dream.

We lack consensus about our place in the Universe. There is no generally agreed upon long-term vision of the goal of our species—other than, perhaps, simple survival. Especially when times are hard, we become desperate for encouragement, unreceptive to the litany of great demotions and dashed hopes, and much more willing to hear that we're special, never mind if the evidence is paper-thin. If it takes a little myth and ritual to get us through a night that seems endless, who among us cannot sympathize and understand?

But if our objective is deep knowledge rather than shallow reassurance, the gains from this new perspective far outweigh the losses. Once we overcome our fear of being tiny, we find ourselves on the threshold of a vast and awesome Universe that utterly dwarfs—in time, in space, and in potential—the tidy anthropocentric proscenium of our ancestors. We gaze across billions of light-years of space to view the Universe shortly after the Big Bang, and plumb the fine structure of matter. We peer down into the core of our planet, and the blazing interior of our star. We read the genetic language in which is written the diverse skills and propensities of every being on Earth. We uncover hidden chapters in the record of our own origins, and with some anguish better understand our nature and prospects. We invent and refine agriculture, without

ABOVE AND OPPOSITE PAGE:

Order in Nature. Does the exquisite regularity of planetary ring systems (left) or spiral galaxies (right) indicate direct intervention by a deity who considers order a grace? (Saturn, backlit, seen by *Voyager 2* after passing it on the spacecraft's way to Uranus, courtesy JPL/NASA.)

which almost all of us would starve to death. We create medicines and vaccines that save the lives of billions. We communicate at the speed of light, and whip around the Earth in an hour and a half. We have sent dozens of ships to more than seventy worlds, and four spacecraft to the stars. We are right to rejoice in our accomplishments, to be proud that our species has been able to see so far, and to judge our merit in part by the very science that has so deflated our pretensions.

To our ancestors there was much in Nature to be afraid of—

Messier 100, in the Virgo Cluster of galaxies, about 62 million light-years away.
Courtesy Anglo-Australian Observatory. Photograph by David Malin.

lightning, storms, earthquakes, volcanos, plagues, drought, long winters. Religions arose in part as attempts to propitiate and control, if not much to understand, the disorderly aspect of Nature. The scientific revolution permitted us to glimpse an underlying ordered Universe in which there was a literal harmony of the worlds (Johannes Kepler's phrase). If we understand Nature, there is a prospect of controlling it or at least mitigating the harm it may bring. In this sense, science brought hope.

Most of the great deprovincializing debates were entered into with no thought for their practical implications. Passionate and curious humans wished to understand their actual circumstances, how unique or pedestrian they and their world are, their ultimate origins and destinies, how the Universe works. Surprisingly, some of these debates have yielded the most profound practical benefits. The very method of mathematical reasoning that Isaac Newton introduced to explain the motion of the planets around the Sun has led to most of the technology of our modern world. The Industrial Revolution, for all its shortcomings, is still the global model of how an agricultural nation can emerge from poverty. These debates have bread-and-butter consequences.

It might have been otherwise. It might have been that the balance lay elsewhere, that humans by and large did not want to know about a disquieting Universe, that we were unwilling to permit challenges to the prevailing wisdom. Despite determined resistance in every age, it is very much to our credit that we have allowed ourselves to follow the evidence, to draw conclusions that at first seem daunting: a Universe so much larger and older that our personal and historical experience is dwarfed and humbled, a Universe in which, every day, suns are born and worlds obliterated, a Universe in which humanity, newly arrived, clings to an obscure clod of matter.

How much more satisfying had we been placed in a garden custom-made for us, its other occupants put there for us to use as we saw fit. There is a celebrated story in the Western tradition like this, except that not quite everything was there for us. There was one particular tree of which we were not to partake, a tree of knowledge. Knowledge and understanding and wisdom were forbidden to us in this story. We were to be kept ignorant. But we couldn't help ourselves. We were starving for knowledge—created hungry, you might say. This was the origin of all our troubles. In particular, it is why we no longer live in a garden: We found out too much. So long as we were incurious and obedient, I imagine, we could console ourselves with our importance and centrality, and tell ourselves that we were the reason the Universe was made. As we began to indulge our curiosity, though, to explore, to learn how the Universe really is, we expelled ourselves from Eden. Angels with a flaming sword were set as sentries at the gates of Paradise to bar our return. The gardeners became exiles and wanderers. Occasionally we mourn that lost world, but that, it seems to me, is maudlin and sentimental. We could not happily have remained ignorant forever.

There is in this Universe much of what seems to be design. Every time we come upon it, we breathe a sigh of relief. We are forever hoping to find, or at least safely deduce, a Designer. But instead, we repeatedly discover that natural processes—collisional selection of worlds, say, or natural selection of gene pools, or even the convection pattern in a pot of boiling water—can extract order out of chaos, and deceive us into deducing purpose where there is none. In everyday life, we often sense—in the bedrooms of teenagers, or in national politics—that chaos is natural, and order imposed from above. While there are deeper regularities in the Universe than the simple circumstances we generally describe as orderly, all that order, simple and complex, seems to derive from laws of Nature established at the Big Bang (or earlier), rather than as a consequence of belated intervention by an imperfect deity. "God is to be found in the details" is the famous dictum of the German scholar Aby Warburg. But, amid much elegance and precision, the details of life and the Universe also exhibit haphazard, jury-rigged arrangements and much poor planning. What shall we make of this: an edifice abandoned early in construction by the architect?

The evidence, so far at least and laws of Nature aside, does not require a Designer. Maybe there is one hiding, maddeningly unwilling to be revealed. Sometimes it seems a very slender hope.

The significance of our lives and our fragile planet is then determined only by our own wisdom and courage. *We* are the custodians of life's meaning. We long for a Parent to care for us, to forgive us our errors, to save us from our childish mistakes. But knowledge is preferable to ignorance. Better by far to embrace the hard truth than a reassuring fable.

If we crave some cosmic purpose, then let us find ourselves a worthy goal.

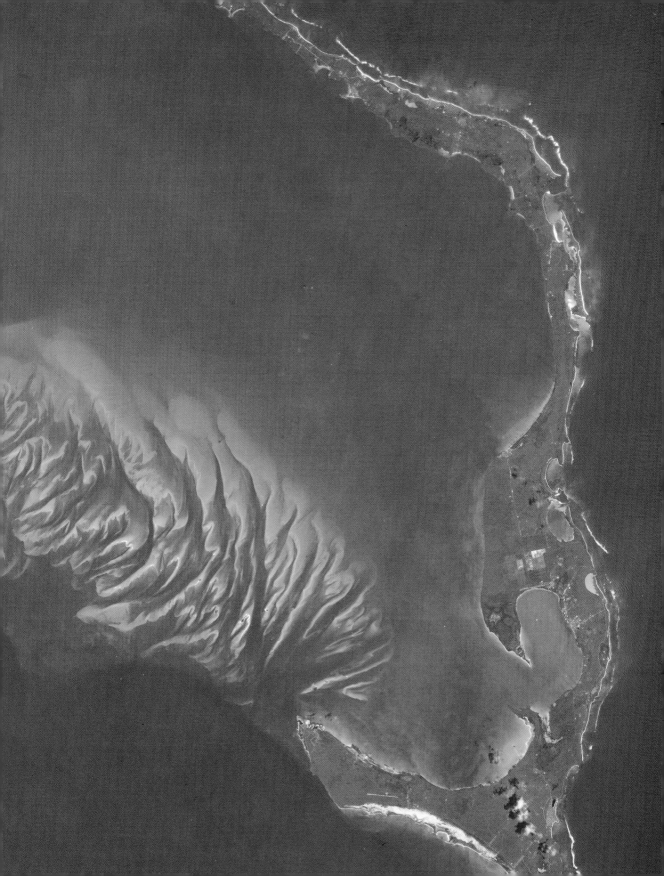

CHAPTER 5

IS THERE INTELLIGENT LIFE ON EARTH?

> They journeyed a long time and found nothing. At length they discerned a small light, which was the Earth . . . [But] they could not find the smallest reason to suspect that we and our fellow-citizens of this globe have the honor to exist.
>
> —VOLTAIRE, *MICROMEGAS.*
> *A PHILOSOPHICAL HISTORY* (1752)

*T*here are places, in and around our great cities, where the natural world has all but disappeared. You can make out streets and sidewalks, autos, parking garages, advertising billboards, monuments of glass and steel, but not a tree or a blade of grass or any animal—besides, of course, the humans. There are lots of humans. Only when you look straight up through the skyscraper canyons can you make out a star or a patch of blue—reminders of what was there long before humans came to be. But the bright lights of the big cities bleach out the stars, and even that patch of blue is sometimes gone, tinted brown by industrial technology.

OPPOSITE: Is there life on Earth? The island of Eleuthera in the Bahamas; false-color imagery by the SPOT satellite. ©1994 by CNES. Provided by SPOT Image Corporation.

It's not hard, going to work every day in such a place, to be impressed with ourselves. How we've transformed the Earth for our benefit and convenience! But a few hundred miles up or down there are no humans. Apart from a thin film of life at the very surface of the Earth, an occasional intrepid spacecraft, and some radio static, our impact on the Universe is nil. It knows nothing of us.

The Earth from the *Galileo* spacecraft. There is no hint of life. Courtesy JPL/NASA.

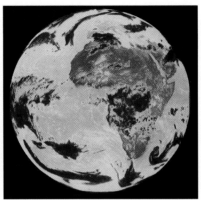

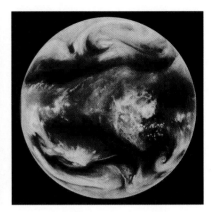

The Earth in reflected visible sunlight observed from space. East Africa, where the human species began, can be clearly seen, but at this resolution there is no sign of life. METEOSAT image, courtesy European Space Agency (ESA).

The Earth from space in emitted infrared radiation. This image was taken at the same moment as the previous picture. Hottest regions are shown in yellow, grading to red, gray, and then blue. On this particular day it was hot in the western Sahara. Dense clouds lie over Zaire and haze over Europe. METEOSAT image, courtesy ESA.

The Earth in reflected near-infrared light—where carbon dioxide and water vapor both absorb light. All the details seen reflect conditions in the upper atmosphere. The ground cannot be made out at all. The difference between this picture and the preceding two is alone sufficient to indicate a substantial greenhouse effect on Earth. METEOSAT image, courtesy ESA.

YOU'RE AN ALIEN EXPLORER entering the Solar System after a long journey through the blackness of interstellar space. You examine the planets of this humdrum star from afar—a pretty handful, some gray, some blue, some red, some yellow. You're interested in what kinds of worlds these are, whether their environments are static or changing, and especially whether there are life and intelligence. You have no prior knowledge of the Earth. You've just discovered its existence.

There's a galactic ethic, let's imagine: Look but don't touch. You can fly by these worlds; you can orbit them; but you are strictly forbidden to land. Under such constraints, could you figure out what the Earth's environment is like and whether anyone lives there?

As you approach, your first impression of the whole Earth is white clouds, white polar caps, brown continents, and some bluish

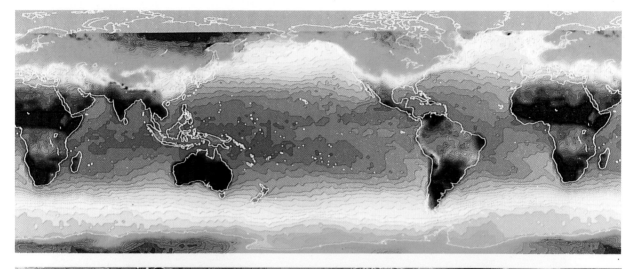

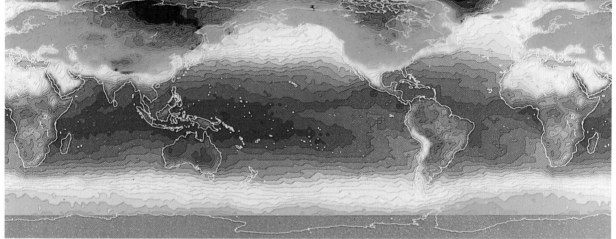

substance that covers two thirds of the surface. When you take the temperature of this world from the infrared radiation it emits, you find that most latitudes are above the freezing point of water, while the polar caps are below freezing. Water is a very abundant material in the Universe; polar caps made of solid water would be a reasonable guess, as well as clouds of solid and liquid water.

You might also be tempted by the idea that the blue stuff is enormous quantities—kilometers deep—of liquid water. The suggestion is bizarre, though, at least as far as *this* solar system is concerned, because surface oceans of liquid water exist nowhere else. When you look in the visible and near-infrared spectrum for telltale signatures of chemical composition, sure enough you discover water ice in the polar caps, and enough water vapor in the air to account for the clouds; this is also just the right amount that must exist because of evaporation if the oceans are in fact made of liquid water. The bizarre hypothesis is confirmed.

The spectrometers further reveal that the air on this world is one fifth oxygen, O_2. No other planet in the Solar System has anything close to so much oxygen. Where does it all come from? The intense ultraviolet light from the Sun breaks water, H_2O, down into oxygen and hydrogen, and hydrogen, the lightest gas, quickly escapes to space. This is a source of O_2, certainly, but it doesn't easily account for so *much* oxygen.

Another possibility is that ordinary visible light, which the Sun pours out in vast amounts, is used on Earth to break water apart—except that there's no known way to do this without life. There would have to be plants—life-forms colored by a pigment that strongly absorbs visible light, that knows how to split a water molecule by saving up the energy of two photons of light, that retains the H and excretes the O, and that uses the hydrogen thus liberated to synthesize organic molecules. The plants would have to be spread over much of the planet. All this is asking a lot. If you're a good skeptical scientist, so much O_2 would not be proof of life. But it certainly might be cause for suspicion.

With all that oxygen you're not surprised to discover ozone (O_3) in the atmosphere, because ultraviolet light makes ozone out of molecular oxygen (O_2). The ozone then absorbs dangerous ul-

OPPOSITE: What are Earth's oceans made of? The average temperature of the Earth (in January 1979). At top, in daytime. Brown is hottest, and the temperatures decline from reds to light blues to dark blues. In this Mercator projection, Africa is shown twice. It is summer in the South. Australia, the southern half of South America, and South Africa are among the hottest places on Earth. At middle are shown average monthly temperatures, but at night. All the places hot in daytime have cooled off. At bottom is the difference between daytime and nighttime temperatures. The Sahara and Australia show the greatest day/night temperature change; this is characteristic of deserts. No temperature difference at all between day and night is seen in the oceans. From this fact alone we would be led to suppose that the oceans are made of liquid water, a substance remarkable for its reluctance to change temperature. NASA satellite data, courtesy Moustafa Chahine, JPL.

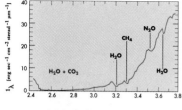

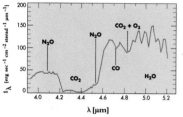

Galileo infrared spectra of the Earth, showing the presence of water vapor, methane, carbon dioxide, carbon monoxide, and nitrous oxide.

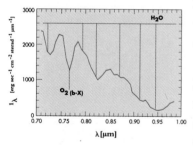

Visible and near-infrared spectra of the Earth obtained by the Galileo spacecraft, showing the presence of water vapor and large quantities of molecular oxygen.

traviolet radiation. So if the oxygen *is* due to life, there's a curious sense in which the life is protecting itself. But this life might be mere photosynthetic plants. A high level of intelligence is not implied.

When you examine the continents more closely, you find there are, crudely speaking, two kinds of regions. One shows the spectrum of ordinary rocks and minerals as found on many worlds. The other reveals something unusual: a material, covering vast areas, that strongly absorbs red light. (The Sun, of course, shines in light of all colors, with a peak in the yellow.) This pigment might be just the agent needed if ordinary visible light *is* being used to break water apart and account for the oxygen in the air. It's another hint, this time a little stronger, of life—not a bug here and there, but a planetary surface overflowing with life. The pigment is in fact chlorophyll: It absorbs blue light as well as red, and is responsible for the fact that plants are green. What you're seeing is a densely vegetated planet.

So the Earth is revealed to possess three properties unique at least in this solar system—oceans, oxygen, life. It's hard not to think they're related, the oceans being the sites of origin, and the oxygen the product, of abundant life.

When you look carefully at the infrared spectrum of the Earth, you discover the minor constituents of the air. In addition to water vapor, there's carbon dioxide (CO_2), methane (CH_4), and other gases that absorb the heat that the Earth tries to radiate away to space at night. These gases warm the planet. Without them, the Earth would everywhere be below the freezing point of water. You've discovered this world's greenhouse effect.

Methane and oxygen together in the same atmosphere is peculiar. The laws of chemistry are very clear: In an excess of O_2, CH_4 should be entirely converted into H_2O and CO_2. The process is so efficient that not a single molecule in all the Earth's atmosphere should be methane. Instead, you find that one out of every million molecules is methane, an immense discrepancy. What could it mean?

The only possible explanation is that methane is being injected into the Earth's atmosphere so quickly that its chemical re-

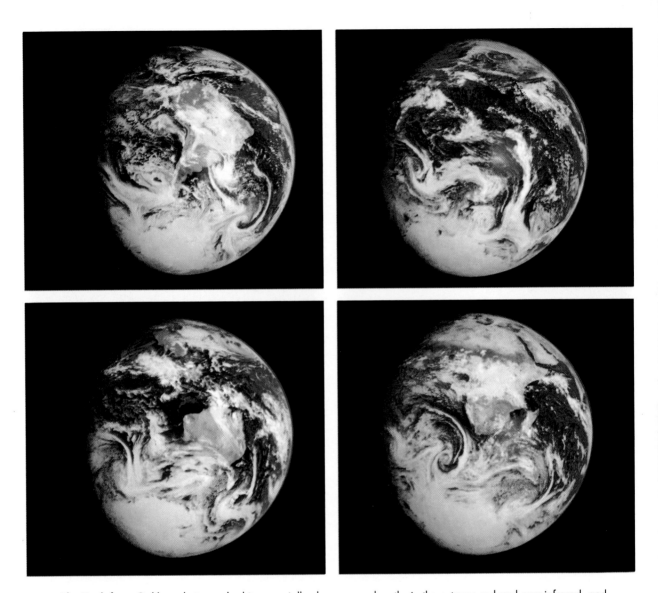

The Earth from *Galileo,* photographed in especially chosen wavelengths in the extreme red and near-infrared, and composited into false-color images. Oxygen in the air shows up as blue. There seems to be more of it toward the poles, because we are looking at a greater slant path through the atmosphere when we look at the poles. Water vapor shows up as magenta and is associated with clouds. Ordinary silicate minerals appear as gray. But the continents of this planet are painted with a pigment that in this false-color composite shows up as orange. No other planet in the Solar System displays this pigment. It is in fact chlorophyll, and a clear indication of life on Earth. From top left, the pictures are centered on South America; the central Pacific; Australia and Indonesia; and Africa. In its December 1990 flyby of the Earth, *Galileo* made its closest approach (only 900 kilometers) over Australia and Antarctica. False-color images prepared by W. Reid Thompson, Cornell University.

action with O_2 can't keep pace. Where does all this methane come from? Maybe it seeps out of the deep interior of the Earth—but quantitatively this doesn't seem to work, and Mars and Venus don't have anything like this much methane. The only alternatives are biological, a conclusion that makes no assumptions about the chemistry of life, or what it looks like, but follows merely from how unstable methane is in an oxygen atmosphere. In fact, the methane arises from such sources as bacteria in bogs, the cultivation of rice, the burning of vegetation, natural gas from oil wells, and bovine flatulence. In an oxygen atmosphere, methane is a sign of life.

That the intimate intestinal activities of cows should be detectable from interplanetary space is a little disconcerting, especially when so much of what we hold dear is not. But an alien scientist flying by the Earth would, at this point, be unable to deduce bogs, rice, fire, oil, or cows. Just life.

All the signs of life that we've discussed so far are due to comparatively simple forms (the methane in the rumens of cows is generated by bacteria that homestead there). Had your spacecraft flown by the Earth a hundred million years ago, in the age of the dinosaurs when there were no humans and no technology, you would still have seen oxygen and ozone, the chlorophyll pigment, and far too much methane. At present, though, your instruments are finding signs not just of life, but of high technology—something that couldn't possibly have been detected even a hundred years ago:

You are detecting a particular kind of radio wave emanating from Earth. Radio waves don't necessarily signify life and intelligence. Many natural processes generate them. You've already found radio emissions from other, apparently uninhabited worlds—generated by electrons trapped in the strong magnetic fields of planets, by chaotic motions at the shock front that separates these magnetic fields from the interplanetary magnetic field, and by lightning. (Radio "whistlers" usually sweep from high notes to low, and then begin again.) Some of these radio emissions are continuous; some come in repetitive bursts; some last a few minutes and then disappear.

But this is different: A portion of the radio transmission from Earth is at just the frequencies where radio waves begin to leak out of the planet's ionosphere, the electrically charged region above the stratosphere that reflects and absorbs radio waves. There is a constant central frequency for each transmission, added to which is a modulated signal (a complex sequence of ons and offs). No electrons in magnetic fields, no shock waves, no lightning discharges can generate something like this. Intelligent life seems to be the only possible explanation. Your conclusion that the radio transmission is due to technology on Earth holds no matter what the ons and offs mean: You don't have to decode the message to be sure it *is* a message. (This signal is really, let us suppose, communications from the U.S. Navy to its distant nuclear-armed submarines.)

So, as an alien explorer, you would know that at least one species on Earth has achieved radio technology. Which one is it? The beings that make methane? Those that generate oxygen? The ones whose pigment colors the landscape green? Or somebody else, somebody more subtle, someone not otherwise detectable to a spacecraft plummeting by? To search for this technological species, you might want to examine the Earth at finer and finer resolution—seeking, if not the beings themselves, at least their artifacts.

You look first with a modest telescope, so the finest detail you can resolve is one or two kilometers across. You can make out no monumental architecture, no strange formations, no unnatural reworking of the landscape, no signs of life. You see a dense atmosphere in motion. The abundant water must evaporate and then rain back down. Ancient impact craters, apparent on the Earth's nearby Moon, are almost wholly absent. There must, then, be a set of processes whereby new land is created and then eroded away in much less time than the age of this world. Running water is implicated. As you look with finer and finer definition you find mountain ranges, river valleys, and many other indications that the planet is geologically active. There are also odd places surrounded by vegetation, but which are themselves denuded of plants. They look like discolored smudges on the landscape.

Western Europe at tens of kilometers resolution. There is still no sign of life. METEOSAT image, courtesy ESA.

Upstate New York in early Landsat false-color imagery. Water is black, clouds are white, and vegetation is red. Lake Ontario is at the top. The Finger Lakes are at the bottom; the two biggest are Seneca Lake (at left) and Cayuga Lake (at right). At the very southern tip of Cayuga Lake is Ithaca, N.Y., where the author lives. In this picture, no sign of life or intelligence is apparent there. Courtesy NASA.

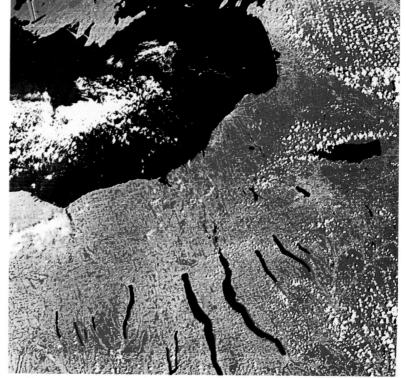

False-color imagery of New York City and vicinity by an early Landsat orbiter. The resolution is about a hundred meters, and New York City—like all cities on Earth—shows as a dark smudge. Courtesy NASA.

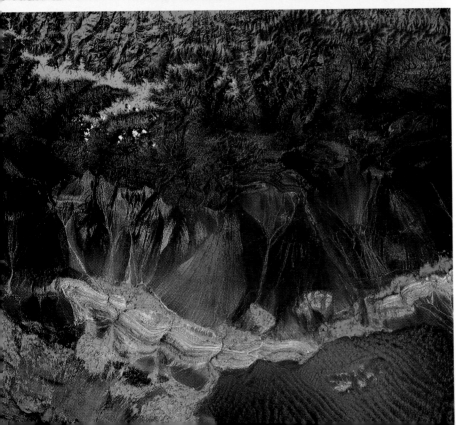

The edge of the Gobi Desert, China, in Landsat false color. Sand dunes can be seen in brown, lower right; snow is in blue, alluvial fans in light purple, and healthy vegetation in green. Reproduced by permission of Earth Observation Satellite Company, Lanham, Maryland, U.S.A.

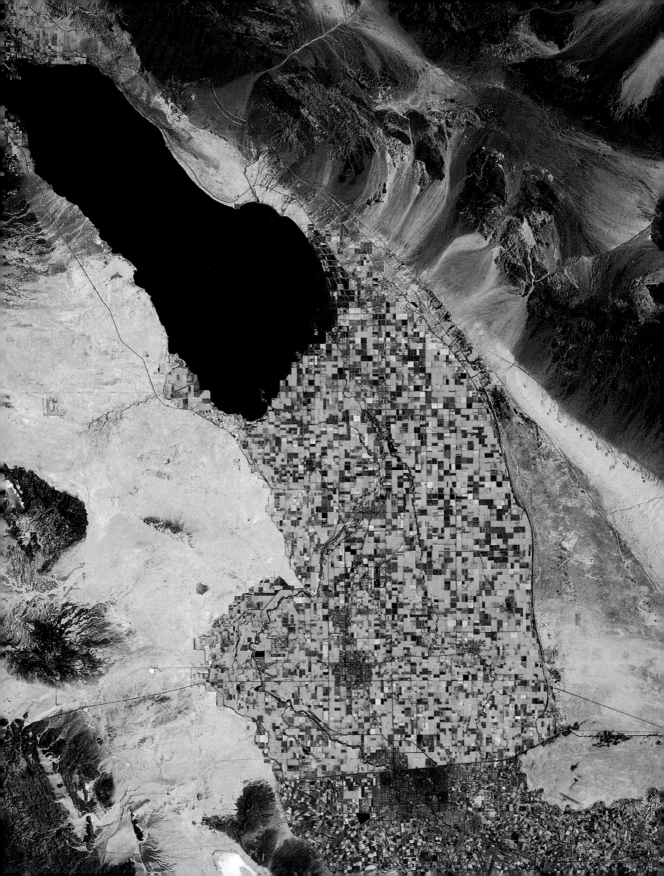

When you examine the Earth at about 100-meter resolution, everything changes. The planet is revealed to be covered with straight lines, squares, rectangles, circles—sometimes huddling along river banks or nestling on the lower slopes of mountains, sometimes stretching over plains, but rarely in deserts or high mountains, and absolutely never in the oceans. Their regularity, complexity, and distribution would be hard to explain except by life and intelligence, although a deeper understanding of function and purpose might be elusive. Perhaps you would conclude only that the dominant life-forms have a simultaneous passion for territoriality and Euclidean geometry. At this resolution you could not see them, much less know them.

Many of the devegetated smudges are revealed to have an underlying checkerboard geometry. These are the planet's cities. Over much of the landscape, and not just in the cities, there is a profusion of straight lines, squares, rectangles, circles. The dark smudges of the cities are revealed to be highly geometrized, with only a few patches of vegetation—themselves with highly regular boundaries—left intact. There are occasional triangles, and in one city there is even a pentagon.

OPPOSITE: A sign of life on Earth: In this false-color image, the black oblong object is the Salton Sea in southern California. The green checkerboard pattern is due to vegetation—agricultural fields organized in neat squares and rectangles. The horizontal boundary near the bottom in the checkerboard pattern is the border between the United States (above) and Mexico (below). Landsat photo. Reproduced by permission of Earth Observation Satellite Company, Lanham, Maryland, U.S.A.

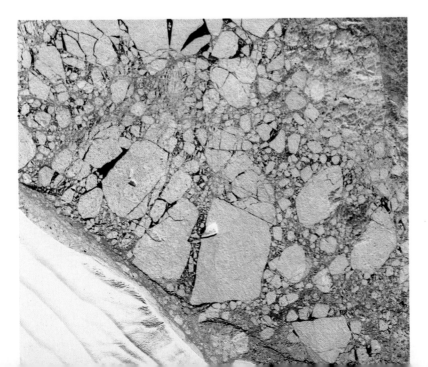

Glacial ice in the Weddell Sea, Antarctica, at a resolution of about 10 meters. There are straight lines but no signs of life. ©1994 by CNES. Provided by SPOT Image Corporation.

Saudi Arabia in Landsat false color: vegetation in the desert. The circular features are agricultural fields watered by center-pivot irrigation systems— newly seeded fields in black, healthy unharvested vegetation in red. Reproduced by permission of Earth Observation Satellite Company, Lanham, Maryland, U.S.A.

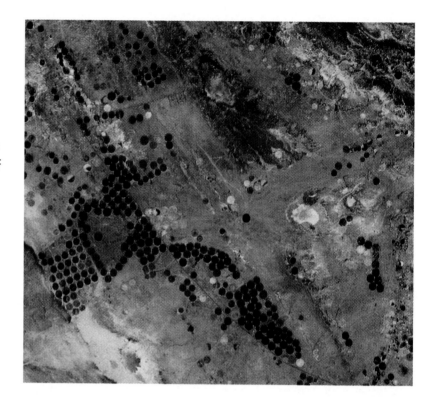

OPPOSITE: New York City and vicinity from the SPOT spacecraft at a resolution of around 10 meters. The regular geometry of streets, highways, bridges, and wharfs is apparent. Note the perfect rectangular shape of the still-vegetated Central Park on Manhattan Island. ©1994 by CNES. Provided by SPOT Image Corporation.

When you take pictures at a meter resolution or better, you find that the crisscrossing straight lines within the cities and the long straight lines that join them with other cities are filled with streamlined, multicolored beings a few meters in length, politely running one behind the other, in long, slow orderly procession. They are very patient. One stream of beings stops so another stream can continue at right angles. Periodically, the favor is returned. At night, they turn on two bright lights in front so they can see where they're going. Some, a privileged few, go into little houses when their workday is done and retire for the night. Most are homeless and sleep in the streets.

At last! You've detected the source of all the technology, the dominant life-forms on the planet. The streets of the cities and the roadways of the countryside are evidently built for their benefit. You might believe that you were really beginning to understand life on Earth. And perhaps you'd be right.

If the resolution improved just a little further, you'd discover

Paris. The scene is bisected by
the snaking River Seine. Close
inspection shows many bridges
across the Seine and all the major
streets. The Arc de Triomphe is in
the middle of the picture, left of
center, at the center of radiating
avenues, including the Champs-
Elysées. ©1994 by CNES.
Provided by SPOT Image
Corporation.

tiny parasites that occasionally enter and exit the dominant organisms. They play some deeper role, though, because a stationary
dominant organism will often start up again just after it's rein-
fected by a parasite, and stop again just before the parasite is expelled. This is puzzling. But no one said life on Earth would be
easy to understand.

All the images you've taken so far are in reflected sunlight—
that is, on the day side of the planet. Something most interesting is
revealed when you photograph the Earth at night: The planet is lit
up. The brightest region, near the Arctic Circle, is illuminated by
the aurora borealis—generated not by life, but by electrons and

protons from the Sun, beamed down by the Earth's magnetic field. Everything else you see is due to life. The lights recognizably outline the same continents you can make out in daytime; and many correspond to cities you've already mapped. The cities are concentrated near coastlines. They tend to be sparser in continental interiors. Perhaps the dominant organisms are desperate for seawater (or maybe oceangoing ships were once essential for commerce and emigration).

Some of the lights, though, are not due to cities. In North Africa, the Middle East, and Siberia, for example, there are very bright lights in a comparatively barren landscape—due, it turns out, to burnoff in oil and natural gas wells. In the Sea of Japan on

Washington, D.C., at high resolution. The U.S. Capitol can be recognized toward upper right, surrounded by a little greenery (shown in false-color red), with many streets radiating from it. Near the bridges across the Potomac (left, middle) can be found, among straight lines, squares, and rectangles, a pentagon. ©1994 by CNES. Provided by SPOT Image Corporation.

The Earth at night in Mercator projection. Image by the Defense Meteorological Satellite. Courtesy Woodruff Sullivan, University of Washington, and U.S. Department of Defense.

the day you first look, there is a strange, triangular-shaped area of light. In daytime it corresponds to open ocean. This is no city. What could it be? It is in fact the Japanese squid fishing fleet, using brilliant illumination to attract schools of squid up to their deaths. On other days, this pattern of light wanders all over the Pacific Ocean, seeking prey. In effect, what you have discovered here is sushi.

It seems sobering to me that from space you can so readily detect some of the odds and ends of life on Earth—the gastrointestinal habits of ruminants, Japanese cuisine, the means of communicating with nomadic submarines that carry death for 200 cities—while so much of our monumental architecture, our greatest engineering works, our efforts to care for one another, are almost wholly invisible. It's a kind of parable.

BY THIS POINT your expedition to the Earth must be considered highly successful. You've characterized the environment; you've detected life; you've found manifestations of intelligent beings; you may even have identified the dominant species, the one transfixed with geometry and rectilinearity. Surely this planet is worth a longer and more detailed study. That's why you've now inserted your spacecraft into orbit around the Earth.

Looking down on the planet, you uncover new puzzles. All over the Earth, smokestacks are pouring carbon dioxide and toxic

The Amazon rain forest at twilight. Every bright point of light is a forest fire. The white clouds are the resulting smoke. ©1994 by CNES. Provided by SPOT Image Corporation.

Spacecraft from the Earth have now flown by dozens of planets, moons, comets, and asteroids—equipped with cameras, instruments for measuring heat and radio waves, spectrometers to determine composition, and a host of other devices. We have found not a hint of life anywhere else in the Solar System. But you might be skeptical about our ability to detect life elsewhere, especially life different from the kind we know. Until recently we had never performed the obvious calibration test: to fly a modern interplanetary spacecraft by the Earth and see whether we could detect ourselves. This all changed on December 8, 1990.

Galileo is a NASA spacecraft designed to explore the giant planet Jupiter, its moons, and its rings. It's named after the heroic Italian scientist who played so central a role in toppling the geocentric pretension. It is he who first saw Jupiter as a world, and who discovered its four big moons. To get to Jupiter, the spacecraft had to fly close by Venus (once) and the Earth (twice) and be accelerated by the gravities of these planets—otherwise there wasn't enough oomph to get it where it was going. This necessity of trajectory design permitted us, for the first time, to look systematically at the Earth from an alien perspective.

Galileo passed only 960 kilometers (about 600 miles) above the Earth's surface. With some exceptions—including pictures showing features finer than 1 kilometer across, and the images of the Earth at night—much of the spacecraft data described in this chapter were actually obtained by *Galileo*. With *Galileo* we were able to deduce an oxygen atmosphere, water, clouds, oceans, polar ice, life, and intelligence. The use of instruments and protocols developed to explore the planets to monitor the environmental health of our own—something NASA is now doing in earnest—was described by the astronaut Sally Ride as "Mission to Planet Earth."

Other members of the NASA scientific team who worked with me on *Galileo*'s detection of life on Earth were Drs. W. Reid Thompson, Cornell University; Robert Carlson, JPL; Donald Gurnett, University of Iowa; and Charles Hord, University of Colorado.

Our success in detecting life on Earth with *Galileo,* without making any assumptions beforehand about what kind of life it must be, increases our confidence that when we fail to find life on other planets, that negative result is meaningful. Is this judgment anthropocentric, geocentric, provincial? I don't think so. We're not looking only for our kind of biology. Any widespread photosynthetic pigment, any gas grossly out of equilibrium with the rest of the atmosphere, any rendering of the surface into highly geometrized patterns, any steady constellation of lights on the night hemisphere, any nonastrophysical sources of radio emission would betoken the presence of life. On Earth we have found of course only our type, but many other types would have been detectable elsewhere. We have not found them. This examination of the third planet strengthens our tentative conclusion that, of all the worlds in the Solar System, only ours is graced by life.

We have just begun to search. Maybe life is hiding on Mars or Jupiter, Europa or Titan. Maybe the Galaxy is filled with worlds as rich in life as ours. Maybe we are on the verge of making such discoveries. But in terms of actual knowledge, at this moment the Earth *is* unique. No other world is yet known to harbor even a microbe, much less a technical civilization.

chemicals into the air. So are the dominant beings who run on the roadways. But carbon dioxide is a greenhouse gas. As you watch, the amount of it in the atmosphere increases steadily, year after year. The same is true of methane and other greenhouse gases. If this keeps up, the temperature of the planet is going to increase. Spectroscopically, you discover another class of molecules being injected into the air, the chlorofluorocarbons. Not only are they greenhouse gases, but they are also devastatingly effective in destroying the protective ozone layer.

You look more closely at the center of the South American continent, which—as you know by now—is a vast rain forest. Every night you see thousands of fires. In the daytime, you find the region covered with smoke. Over the years, all over the planet, you find less and less forest and more and more scrub desert.

You look down on the large island of Madagascar. The rivers are colored brown, generating a vast stain in the surrounding ocean. This is topsoil being washed out to sea at a rate so high that in another few decades there will be none left. The same thing is happening, you note, at the mouths of rivers all over the planet.

But no topsoil means no agriculture. In another century, what will they eat? What will they breathe? How will they cope with a changing and more dangerous environment?

From your orbital perspective, you can see that something has unmistakably gone wrong. The dominant organisms, whoever they are—who have gone to so much trouble to rework the surface— are simultaneously destroying their ozone layer and their forests, eroding their topsoil, and performing massive, uncontrolled experiments on their planet's climate. Haven't they noticed what's happening? Are they oblivious to their fate? Are they unable to work together on behalf of the environment that sustains them all?

Perhaps, you think, it's time to reassess the conjecture that there's intelligent life on Earth.

The *Galileo* spacecraft emerges from the cargo bay of the *Atlantis* space shuttle on its way to the Main Belt asteroids Gaspra and Ida, to Jupiter, and (incidentally) to Venus and the Earth. Courtesy JPL/NASA.

THE TRIUMPH
OF *VOYAGER*

They that go down to the sea in ships, that do business in great waters;
these see the works of the Lord, and his wonders in the deep.
—PSALMS, 107 (CA. 150 B.C.)

The visions we offer our children shape the future. It *matters* what those visions are. Often they become self-fulfilling prophecies. Dreams are maps.

I do not think it irresponsible to portray even the direst futures; if we are to avoid them, we must understand that they are possible. But where are the alternatives? Where are the dreams that motivate and inspire? We long for realistic maps of a world we can be proud to give to our children. Where are the cartographers of human purpose? Where are the visions of hopeful futures, of technology as a tool for human betterment and not a gun on hair trigger pointed at our heads?

NASA, in its ordinary course of doing business, offers such a vision. But in the 1980s and early '90s, many people saw the U.S. space program as, instead, a succession of catastrophes—seven

OPPOSITE: Saturn as photographed by *Voyager 1* on November 3, 1980, from a distance of 13 million kilometers (8 million miles). The major break in the rings is called the Cassini Division, after the seventeenth-century Italian-French astronomer J. D. Cassini. The two moons of Saturn shown are Tethys (above) and Dione (below). Shadows of Saturn's rings and of the moon Tethys are cast on the clouds of Saturn. Courtesy JPL/NASA.

brave Americans killed on a mission whose main function was to put up a communications satellite that could have been launched at less cost without risking anybody; a billion-dollar telescope sent up with a bad case of myopia; a spacecraft to Jupiter whose main antenna—essential for returning data to Earth—did not unfurl; a probe lost just as it was about to orbit Mars. Some people cringe every time NASA describes as exploration sending a few astronauts 200 miles up in a small capsule that endlessly circles the Earth and goes nowhere. Compared to the brilliant achievements of robotic missions, it is striking how rarely fundamental scientific findings emerge from manned missions. Except for repairing ineptly manufactured or malfunctioning satellites, or launching a satellite that could just as well have been sent up in an unmanned booster, the manned program has, since the 1970s, seemed unable to generate accomplishments commensurate with the cost. Others looked at NASA as a stalking horse for grandiose schemes to put weapons into space, despite the fact that an orbiting weapon is in many circumstances a sitting duck. And NASA showed many symptoms of an aging, arteriosclerotic, overcautious, unadventurous bureaucracy. The trend is perhaps beginning to be reversed.

But these criticisms—many of them surely valid—should not blind us to NASA triumphs in the same period: the first exploration of the Uranus and Neptune systems, the in-orbit repair of the Hubble space telescope, the proof that the existence of galaxies is compatible with the Big Bang, the first close-up observations of asteroids, mapping Venus pole to pole, monitoring ozone depletion, demonstrating the existence of a black hole with the mass of a billion suns at the center of a nearby galaxy, and a historic commitment to joint space endeavors by the U.S. and Russia.

There are far-reaching, visionary, and even revolutionary implications to the space program. Communications satellites link up the planet, are central to the global economy, and, through television, routinely convey the essential fact that we live in a global community. Meteorological satellites predict the weather, save lives in hurricanes and tornados, and avoid many billions of dollars in crop losses every year. Military-reconnaissance and treaty-verification satellites make nations and the global civilization more secure;

in a world with tens of thousands of nuclear weapons, they calm the hotheads and paranoids on all sides; they are essential tools for survival on a troubled and unpredictable planet.

Earth-observing satellites, especially a new generation soon to be deployed, monitor the health of the global environment: greenhouse warming, topsoil erosion, ozone layer depletion, ocean currents, acid rain, the effects of floods and droughts, and new dangers we haven't yet discovered. This is straightforward planetary hygiene.

Global positioning systems are now in place so that your locale is radio-triangulated by several satellites. Holding a small instrument the size of a modern shortwave radio, you can read out to high precision your latitude and longitude. No crashed airplane, no ship in fog and shoals, no driver in an unfamiliar city need ever be lost again.

Astronomical satellites peering outward from Earth orbit observe with unsurpassed clarity—studying questions ranging from the possible existence of planets around nearby stars to the origin and fate of the Universe. Planetary probes from close range explore the gorgeous array of other worlds in our solar system, comparing their fates with ours.

All of these activities are forward-looking, hopeful, stirring, and cost-effective. None of them requires "manned"* spaceflight. A key issue facing the future of NASA and addressed in this book is whether the purported justifications for human spaceflight are coherent and sustainable. Is it worth the cost?

But first, let's consider the visions of a hopeful future vouchsafed by robot spacecraft out among the planets.

---

* Since women astronauts and cosmonauts of several nations have flown in space, "manned" is just flat-out incorrect. I've attempted to find an alternative to this widely used term, coined in a more unself-consciously sexist age. I tried "crewed" for a while, but in spoken language it lends itself to misunderstanding. "Piloted" doesn't work, because even commercial airplanes have robot pilots. "Manned and womanned" is just, but unwieldy. Perhaps the best compromise is "human," which permits us to distinguish crisply between human and robotic missions. But every now and then, I find "human" not quite working, and to my dismay "manned" slips back in.

VOYAGER 1 AND VOYAGER 2 are the ships that opened the Solar System for the human species, trailblazing a path for future generations. Before their launch, in August and September 1977, we were almost wholly ignorant about most of the planetary part of the Solar System. In the next dozen years, they provided our first detailed, close-up information on many new worlds—some of them previously known only as fuzzy disks in the eyepieces of ground-based telescopes, some merely as points of light, and some whose very existence was unsuspected. They are still returning reams of data.

These spacecraft have taught us about the wonders of other worlds, about the uniqueness and fragility of our own, about beginnings and ends. They have given us access to most of the Solar System—both in extent and in mass. They are the ships that first explored what may be homelands of our remote descendants.

U.S. launch vehicles are these days too feeble to get such a spacecraft to Jupiter and beyond in only a few years by rocket propulsion alone. But if we're clever (and lucky), there's something else we can do: We can (as *Galileo* also did, years later) fly close to one world, and have its gravity fling us on to the next. A gravity assist, it's called. It costs us almost nothing but ingenuity. It's something like grabbing hold of a post on a moving merry-go-round as it passes—to speed you up and fling you in some new direction. The spacecraft's acceleration is compensated for by a deceleration in the planet's orbital motion around the Sun. But because the planet is so massive compared to the spacecraft, it slows down hardly at all. Each *Voyager* spacecraft picked up a velocity boost of nearly 40,000 miles per hour from Jupiter's gravity. Jupiter in turn was slowed down in its motion around the Sun. By how much? Five billion years from now, when our Sun becomes a swollen red giant, Jupiter will be one millimeter short of where it would have been had *Voyager* not flown by it in the late twentieth century.

Voyager 2 took advantage of a rare lining-up of the planets: A close flyby of Jupiter accelerated it on to Saturn, Saturn to Uranus, Uranus to Neptune, and Neptune to the stars. But you can't do this anytime you like: The previous opportunity for such a game of celestial billiards presented itself during the presidency of

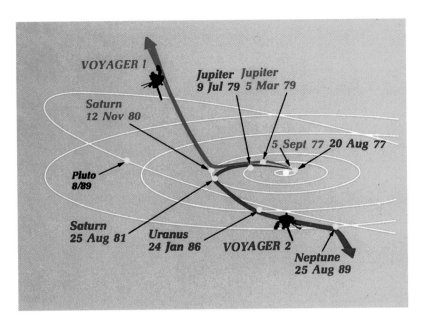

Trajectories of the two *Voyager* spacecraft. Courtesy JPL/NASA.

Thomas Jefferson. We were then only at the horseback, canoe, and sailing ship stage of exploration. (Steamboats were the transforming new technology just around the corner.)

Since adequate funds were unavailable, NASA's Jet Propulsion Laboratory (JPL) could afford to build spacecraft that would work reliably only as far as Saturn. Beyond that, all bets were off. However, because of the brilliance of the engineering design—and the fact that the JPL engineers who radioed instructions up to the spacecraft got smarter faster than the spacecraft got stupid—both spacecraft went on to explore Uranus and Neptune. These days they are broadcasting back discoveries from beyond the most distant known planet of the Sun.

We tend to hear much more about the splendors returned than the ships that brought them, or the shipwrights. It has always been that way. Even those history books enamored of the voyages of Christopher Columbus do not tell us much about the builders of the *Niña,* the *Pinta,* and the *Santa María,* or about the principle of the caravel. These spacecraft, their designers, builders, navigators, and controllers are examples of what science and engineering, set free for well-defined peaceful purposes, can accomplish. Those scientists and engineers should be role models for an

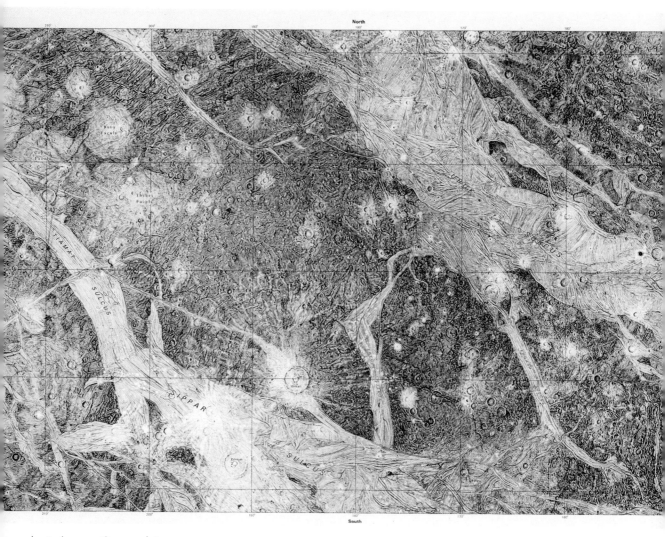

Jupiter's moon Ganymede's equatorial regions, as revealed by *Voyagers* 1 and 2. Many of the features are named after ancient Sumerian cities and gods. USGS shaded relief map.

America seeking excellence and international competitiveness. They should be on our stamps.

At each of the four giant planets—Jupiter, Saturn, Uranus, and Neptune—one or both spacecraft studied the planet itself, its rings, and its moons. At Jupiter, in 1979, they braved a dose of trapped charged particles a thousand times more intense than what it takes to kill a human; enveloped in all that radiation, they discovered the rings of the largest planet, the first active volcanos outside Earth, and a possible underground ocean on an airless world—among a host of surprising discoveries. At Saturn, in 1980 and 1981, they survived a blizzard of ice and found not a few new rings, but thousands. They examined frozen moons mysteriously melted in the comparatively recent past, and a large world with a putative ocean of liquid hydrocarbons surmounted by clouds of organic matter.

On January 25, 1986, *Voyager 2* entered the Uranus system and reported a procession of wonders. The encounter lasted only a few hours, but the data faithfully relayed back to Earth have revolutionized our knowledge of the aquamarine planet, its 15 moons, its pitch-black rings, and its belt of trapped high-energy charged particles. On August 25, 1989, *Voyager 2* swept through the Neptune system and observed, dimly illuminated by the distant Sun, kaleidoscopic cloud patterns and a bizarre moon on which plumes of fine organic particles were being blown about by the astonishingly thin air. And in 1992, having flown beyond the outermost known planet, both *Voyager*s picked up radio emission thought to emanate from the still remote heliopause—the place where the wind from the Sun gives way to the wind from the stars.

Because we're stuck on Earth, we're forced to peer at distant

Exquisite cloud patterns on Jupiter as seen by *Voyager*. Courtesy JPL/NASA.

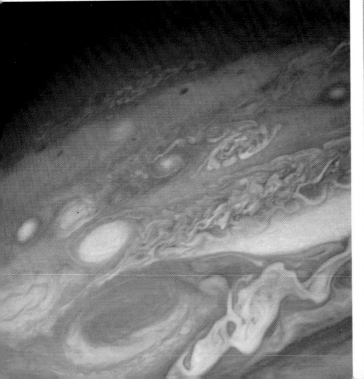

Jupiter as seen from the surface of Europa. The moon at right, surrounded by a torus of escaping gases, is Io. Some scientists think there is an underground ocean of water on Europa. Painting by Don Davis.

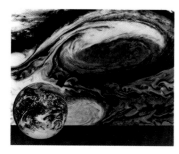

Close-up of Jupiter's Great Red Spot in false color, compared in size to the Earth. Courtesy S.P. Meszaros and NASA.

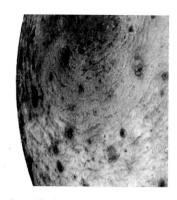

The Valhalla multiringed impact basin on Callisto, in this negative image. Courtesy USGS/NASA.

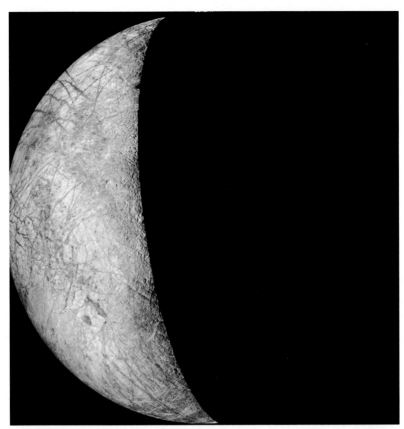

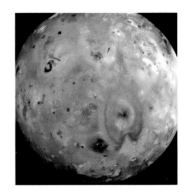

LEFT: Photomosaic of a portion of the surface of Europa as seen by *Voyager*. Courtesy USGS/NASA.

Volcanic Io. *Voyager* photomosaic, courtesy USGS/NASA.

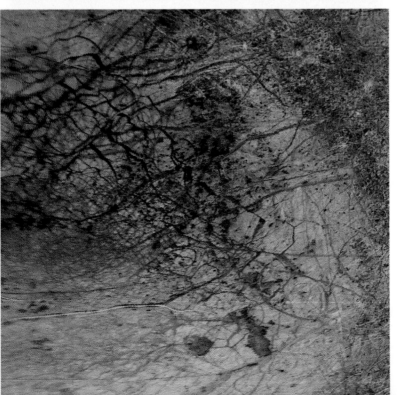

Europa in a high-resolution *Voyager* photomosaic. Courtesy USGS/NASA.

The many rings of Saturn in greatly exaggerated false color, with the Earth inserted to scale for comparison purposes. *Voyager* image, courtesy S.P. Meszaros and NASA.

Saturn in greatly exaggerated false color as seen by *Voyager.* Courtesy JPL/NASA.

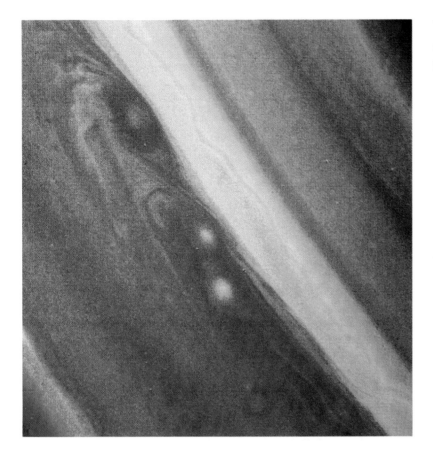

The clouds of Saturn in greatly exaggerated false color. *Voyager* image, courtesy JPL/NASA.

worlds through an ocean of distorting air. Much of the ultraviolet, infrared, and radio waves they emit do not penetrate our atmosphere. It's easy to see why our spacecraft have revolutionized the study of the Solar System: We ascend to stark clarity in the vacuum of space, and there approach our objectives, flying past them, as did *Voyager,* or orbiting them, or landing on their surfaces.

These spacecraft have returned four trillion bits of information to Earth, the equivalent of about 100,000 encyclopedia volumes. I described the *Voyager 1* and *2* encounters with the Jupiter system in *Cosmos.* In the following pages, I'll say something about the Saturn, Uranus, and Neptune encounters.

JUST BEFORE *VOYAGER 2* was to encounter the Uranus system, the mission design had specified a final maneuver, a brief firing of the on-board propulsion system to position the spacecraft correctly so it could thread its way on a preset path among the hurtling moons. But the course correction proved unnecessary. The spacecraft was already within 200 kilometers of its designed trajectory—after a journey along an arcing path 5 billion kilometers long. This is roughly the equivalent of throwing a pin through the eye of a needle 50 kilometers away, or firing your rifle in Washington and hitting the bull's-eye in Dallas.

Mother lodes of planetary treasure were radioed back to Earth. But Earth is so far away that by the time the signal from Neptune was gathered in by radio telescopes on our planet, the received power was only 10^{-16} watts (fifteen zeros between the decimal point and the one). This weak signal bears the same proportion to the power emitted by an ordinary reading lamp as the diameter of an atom bears to the distance from the Earth to the Moon. It's like hearing an amoeba's footstep.

The mission was conceived during the late 1960s. It was first funded in 1972. But it was not approved in its final form (including the encounters with Uranus and Neptune) until after the ships had completed their reconnaissance of Jupiter. The two spacecraft were lifted off the Earth by a nonreusable *Titan/Centaur* booster configuration. Weighing about a ton, a *Voyager* would fill a small house. Each draws about 400 watts of power—considerably less than an average American home—from a generator that converts radioactive plutonium into electricity. (If it had to rely on solar energy, the available power would diminish quickly as the ship ventured farther and farther from the Sun. Were it not for nuclear power, *Voyager* would have returned no data at all from the outer Solar System, except perhaps a little from Jupiter.)

The flow of electricity through the innards of the spacecraft would generate enough magnetism to overwhelm the sensitive instrument that measures interplanetary magnetic fields. So the magnetometer is placed at the end of a long boom, far from the offending electrical currents. With other projections, it gives *Voyager* a slightly porcupine appearance. Cameras, infrared and ultraviolet spectrometers, and an instrument called a photopolarimeter are on a scan platform that swivels on command so these devices can be

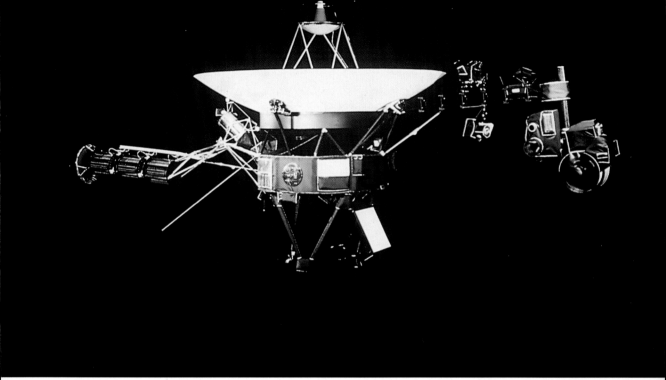

aimed at a target world. The spacecraft must know where Earth is if the antenna is to be pointed properly and the data received back home. It also needs to know where the Sun is and at least one bright star, so it can orient itself in three dimensions and point properly toward any passing world. If you can't point the cameras, it does no good to be able to return pictures over billions of miles.

Each spacecraft cost about as much as a single modern strategic bomber. But unlike bombers, *Voyager* cannot, once launched, be returned to the hangar for repairs. The ship's computers and electronics are therefore designed redundantly. Much key machinery, including the essential radio receiver, had at least one backup—waiting to be called upon should the hour of need ever arrive. When either *Voyager* finds itself in trouble, the computers use branched contingency tree logic to work out the appropriate

The *Voyager* spacecraft. The scan platform containing cameras and spectrometers is at the extreme left. Particle and field detectors are on the various other projections. The antenna for radioing data home and receiving commands from Earth is white and dish-shaped at top. Computers, tape recorders, thermal control devices, and other instrumentation are in the octagonal structure amidships, on one face of which is the *Voyager* Interstellar Record. Courtesy JPL/NASA.

course of action. If that doesn't work, the ship radios home for help.

As the spacecraft journeys increasingly far from Earth, the round-trip radio travel time also increases, approaching eleven hours by the time *Voyager* is at the distance of Neptune. Thus, in case of emergency, the spacecraft needs to know how to put itself into a safe standby mode while awaiting instructions from Earth. As it ages, more and more failures are expected, both in its mechanical parts and in its computer system, although there is no sign, even now, of a serious memory deterioration, some robotic Alzheimer's disease.

This is not to say that *Voyager* is perfect. Serious mission-threatening, white-knuckle mishaps did occur. Each time, special teams of engineers—some of whom had been with the *Voyager* program since its inception—were assigned to "work" the problem. They would study the underlying science and draw upon their previous experience with the failed subsystems. They would experiment with identical *Voyager* spacecraft equipment that had never been launched, or even manufacture a large number of components of the sort that failed in order to gain some statistical understanding of the failure mode.

In April 1978, almost eight months after launch, and while the ship was approaching the asteroid belt, an omitted ground command—a human error—caused *Voyager 2*'s on-board computer to switch from the prime radio receiver to its backup. During the next ground transmission to the spacecraft, the backup receiver refused to lock onto the signal from Earth. A component called a tracking loop capacitor had failed. After seven days in which *Voyager 2* was entirely out of contact, its fault protection software suddenly commanded the backup receiver to be switched off and the prime receiver to be switched back on. Mysteriously—to this day, no one knows why—the prime receiver failed moments later. It was never heard from again. To top it off, the on-board computer now foolishly insisted on using the failed primary receiver. Through an unlucky concatenation of human and robotic error, the spacecraft was now in real jeopardy. No one could think of a way to get *Voyager 2* to revert to the backup receiver. Even if it did, the backup receiver couldn't receive the commands from Earth—because of that failed capacitor. There were many project personnel who feared that all was lost.

But after a week of obdurate unresponsiveness to all commands, instructions to switch automatically between receivers were accepted and programmed into the skittish on-board computer. During that same week the JPL engineers designed an innovative command frequency control procedure to make sure that essential orders would be understood by the damaged backup receiver.

The engineers were now able to recommunicate, at least in a rudimentary way, with the spacecraft. Unfortunately the backup receiver now turned giddy, becoming extremely sensitive to the stray heat dumped when various components of the spacecraft powered up or down. In the following months the JPL engineers devised and conducted tests that let them thoroughly understand the thermal implications of most spacecraft operational modes: What would prevent and what would permit receipt of commands from Earth?

With this information, the backup receiver problem was entirely circumvented. It subsequently acquired all the commands sent from Earth on how to gather data in the Jupiter, Saturn, Uranus, and Neptune systems. The engineers had saved the mission. (To be on the safe side, during most of *Voyager 2*'s subsequent flight a nominal data-taking sequence for the next planet to be encountered was always sitting in the on-board computers—should the spacecraft again become deaf to entreaties from home.)

Another heart-wrenching failure occurred just after *Voyager 2* emerged from behind Saturn (as seen from the Earth) in August 1981. The scan platform had been moving feverishly—pointing here and there among the rings, moons, and the planet itself during the all-too-brief moments of close approach. Suddenly, the platform jammed. A stuck scan platform is a maddening predicament: knowing that the spacecraft is flying past wonders that have never been witnessed, that we will not see again for years or decades, and the incurious spacecraft staring fixedly off into space, ignoring everything.

The scan platform is driven by actuators containing gear trains. So first the JPL engineers ran an identical copy of a flight actuator in a simulated mission. This actuator failed after 348 turns; the actuator on the spacecraft had failed after 352 turns. The problem turned out to be a lubrication failure. Good to know, but what to do about it? Plainly, it would be impossible to overtake *Voyager* with an oilcan.

The engineers wondered whether they could restart the failed actuator by alternate heating and cooling; maybe the resulting thermal stresses would induce the components of the actuator to expand and contract at different rates and unjam the system. They tested this notion with specially manufactured actuators in the laboratory, and then jubilantly found that in this way they could start the scan platform up again in space. Project personnel also devised ways to diagnose any additional trend toward actuator failure early enough to work around the problem. Thereafter, *Voyager 2*'s scan platform worked perfectly. All the pictures taken in the Uranus and Neptune systems owe their existence to this work. The engineers had saved the day again.

*Voyager*s *1* and *2* were designed to explore the Jupiter and Saturn systems only. It is true that their trajectories would carry them on past Uranus and Neptune, but officially these planets

The Mission Operations Facility at JPL during *Voyager* cruise phase (post–Saturn encounter). Courtesy JPL/NASA.

were never contemplated as targets for *Voyager* exploration: The spacecraft were not supposed to last that long. Because of our wish to fly close to the mystery world Titan, *Voyager 1* was flung by Saturn on a path that could never encounter any other known world; it is *Voyager 2* that flew on to Uranus and Neptune with brilliant success. At these immense distances, sunlight is getting progressively dimmer, and the radio signals transmitted to Earth are getting progressively fainter. These were predictable but still very serious problems that the JPL engineers and scientists also had to solve.

Because of the low light levels at Uranus and Neptune, the *Voyager* television cameras were obliged to take long time exposures. But the spacecraft was hurtling so fast through, say, the Uranus system (at about 35,000 miles per hour) that the image would have been smeared or blurred. To compensate, the entire spacecraft had to be moved during the time exposures to cancel out the motion, like panning in the direction opposite yours while taking a photograph of a street scene from a moving car. This may sound easy, but it's not: You have to neutralize the most innocent of motions. At zero gravity, the mere start and stop of the on-board tape recorder can jiggle the spacecraft enough to smear the picture.

This problem was solved by sending up commands to the spacecraft's little rocket engines (called thrusters), machines of exquisite sensitivity. With a little puff of gas at the start and stop of each data-taking sequence, the thrusters compensated for the tape-recorder jiggle by turning the entire spacecraft just a little. To deal with the low radio power received at Earth, the engineers devised a new and more efficient way to record and transmit the data, and the radio telescopes on Earth were electronically linked together with others to increase their sensitivity. Overall, the imaging system worked, by many criteria, better at Uranus and Neptune than it did at Saturn or even at Jupiter.

Voyager may not yet be done exploring. There is, of course, a chance that some vital subsystem will fail tomorrow, but as far as the radioactive decay of the plutonium power source is concerned, the two *Voyager* spacecraft should be able to return data to Earth roughly through the year 2015.

Voyager is an intelligent being—part robot, part human. It extends the human senses to far-off worlds. For simple tasks and short-term problems, it relies on its own intelligence; but for more complex tasks and longer-term problems, it turns to the collective intelligence and experience of the JPL engineers. This trend is sure to grow. The *Voyagers* embody the technology of the early 1970s; if spacecraft were designed for such a mission today, they would incorporate stunning advances in artificial intelligence, in miniaturization, in data-processing speed, in the ability to self-diagnose and repair, and in the propensity to learn from experience. They would also be much cheaper.

In the many environments too dangerous for people, on Earth as well as in space, the future belongs to robot-human partnerships that will recognize the two *Voyagers* as antecedents and pioneers. For nuclear accidents, mine disasters, undersea exploration and archaeology, manufacturing, prowling the interiors of volcanos, and household help, to name only a few potential applications, it could make an enormous difference to have a ready corps of smart, mobile, compact, commandable robots that can diagnose and repair their own malfunctions. There are likely to be many more of this tribe in the near future.

It is conventional wisdom now that anything built by the government will be a disaster. But the two *Voyager* spacecraft were built by the government (in partnership with that other bugaboo, academia). They came in at cost, on time, and vastly exceeded their design specifications—as well as the fondest dreams of their makers. Seeking not to control, threaten, wound, or destroy, these elegant machines represent the exploratory part of our nature set free to roam the Solar System and beyond. This kind of technology, the treasures it uncovers freely available to all humans everywhere, has been, over the last few decades, one of the few activities of the United States admired as much by those who abhor many of its policies as by those who agree with it on every issue. *Voyager* cost each American less than a penny a year from launch to Neptune encounter. Missions to the planets are one of those things—and I mean this not just for the United States, but for the human species—that we do best.

The last close encounter picture.
Voyager 2, on its way to the
stars, photographs Neptune and
its extraordinary moon Triton,
both as thin crescents.
Courtesy JPL/NASA.

AMONG THE MOONS OF SATURN

Seat thyself sultanically among the moons of Saturn.
—HERMAN MELVILLE, *MOBY DICK*, CHAPTER 107 (1851)

There is a world, midway in size between the Moon and Mars, where the upper air is rippling with electricity— streaming in from the archetypical ringed planet next door, where the perpetual brown overcast is tinged with an odd burnt orange, and where the very stuff of life falls out of the skies onto the unknown surface below. It is so far away that light takes more than an hour to get there from the Sun. Spacecraft take years. Much about it is still a mystery—including whether it holds great oceans. We know just enough, though, to recognize that within reach may be a place where certain processes are today

OPPOSITE: Early in the next century, the *Huygens* entry probe of the European Space Agency (ESA) descends through the upper clouds of Titan to the unknown surface below. Painting courtesy Hamid Hassan, ESA.

Photomontage of Saturn and a few of its moons. Titan is at top. Courtesy JPL/NASA.

working themselves out that aeons ago on Earth led to the origin of life.

On our own world a long-standing—and in some respects quite promising—experiment has been under way on the evolution of matter. The oldest known fossils are about 3.6 billion years old. Of course, the origin of life had to have happened well before that. But 4.2 or 4.3 billion years ago the Earth was being so ravaged by the final stages of its formation that life could not yet have come into being: Massive collisions were melting the surface, turning the oceans into steam and driving any atmosphere that had accumulated since the last impact off into space. So around 4 billion years ago, there was a fairly narrow window—perhaps only a hundred million years wide—in which our most distant ancestors came to be. Once conditions permitted, life arose fast. Somehow.

The first living things very likely were inept, far less capable than the most humble microbe alive today—perhaps just barely able to make crude copies of themselves. But natural selection, the key process first coherently described by Charles Darwin, is an instrument of such enormous power that from the most modest beginnings there can emerge all the richness and beauty of the biological world.

Those first living things were made of pieces, parts, building blocks which had to come into being on their own—that is, driven by the laws of physics and chemistry on a lifeless Earth. The building blocks of all terrestrial life are called organic molecules, molecules based on carbon. Of the stupendous number of possible organic molecules, very few are used at the heart of life. The two most important classes are the amino acids, the building blocks of proteins, and the nucleotide bases, the building blocks of the nucleic acids.

Just before the origin of life, where did these molecules come from? There are only two possibilities: from the outside or from the inside. We know that vastly more comets and asteroids were hitting the Earth than do so today, that these small worlds are rich storehouses of complex organic molecules, and that some of these molecules escaped being fried on impact. Here I'm describing homemade, not imported, goods: the organic molecules generated in the air and waters of the primitive Earth.

Unfortunately, we don't know very much about the composition of the early air, and organic molecules are far easier to make in some atmospheres than in others. There couldn't have been much oxygen, because oxygen is generated by green plants and there weren't any green plants yet. There was probably more hydrogen, because hydrogen is very abundant in the Universe and escapes from the upper atmosphere of the Earth into space better than any other atom (because it's so light). If we can imagine various possible early atmospheres, we can duplicate them in the laboratory, supply some energy, and see which organic molecules are made and in what amounts. Such experiments have over the years

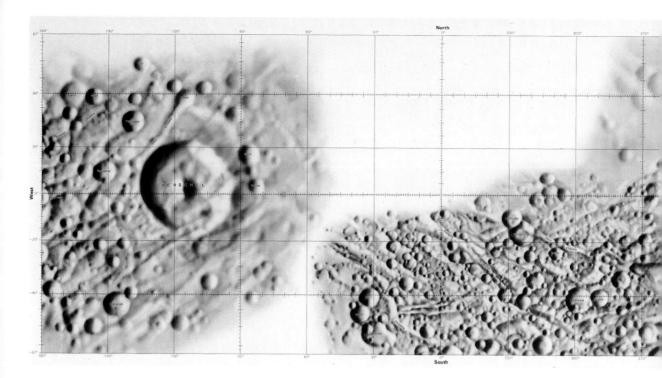

Mimas, the innermost moon of Saturn known before *Voyager*. Finely-rendered features are taken from the highest-resolution *Voyager* pictures. More blurry detail, as the crater Herschel, are from lower-resolution *Voyager* data. The white areas were never imaged and are still unknown lands. Shaded relief map courtesy USGS.

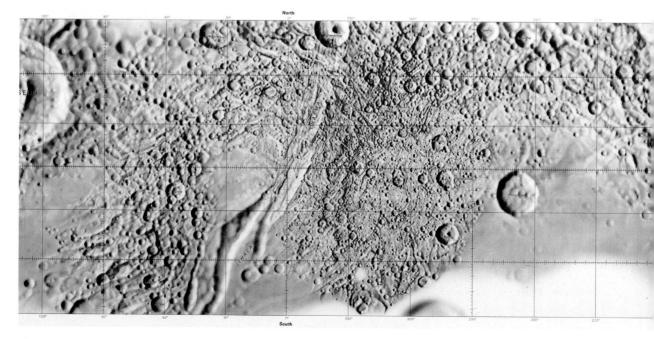

The Saturnian moon Tethys, shaded relief map. Features on this world are named chiefly after characters and locales in Homer's *Odyssey*. The largest impact crater is Odysseus. Ithaca Chasma almost circumnavigates this world. Courtesy USGS.

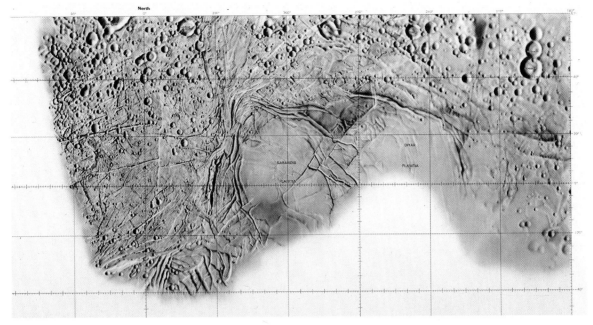

North

Enceladus, an icy moon of
Saturn immersed in one of
Saturn's rings. Note the absence
of impact craters in the southern
provinces, indicating that this
world's surface was recently
melted. No one understands
how this could be. USGS
shaded relief map.

proved provocative and promising. But our ignorance of initial conditions limits their relevance.

What we need is a real world whose atmosphere still retains some of those hydrogen-rich gases, a world in other respects something like the Earth, a world in which the organic building blocks of life are being massively generated in our own time, a world we can go to to seek our own beginnings. There is only one such world in the Solar System.* That world is Titan, the big moon of Saturn. It's about 5,150 kilometers (3,200 miles) in diameter, a little less than half the size of the Earth. It takes 16 of our days to complete one orbit of Saturn.

No world is a perfect replica of any other, and in at least one important respect Titan is very different from the primitive Earth: Being so far from the Sun, its surface is extremely cold, far below the freezing point of water, around 180° below zero Celsius. So while the Earth at the time of the origin of life was, as now, mainly ocean-covered, plainly there can be no oceans of liquid water on Titan. (Oceans made of other stuff are a different story, as we shall see.) The low temperatures provide an advantage, though, because once molecules are synthesized on Titan, they tend to

---

* There could have been none. We're very lucky that there *is* such a world to study. The others all have too much hydrogen, or not enough, or no atmosphere at all.

stick around: The higher the temperature, the faster molecules fall to pieces. On Titan the molecules that have been raining down like manna from heaven for the last 4 billion years might still be there, largely unaltered, deep-frozen, awaiting the chemists from Earth.

THE INVENTION OF THE TELESCOPE in the seventeenth century led to the discovery of many new worlds. In 1610 Galileo first spied the four large satellites of Jupiter. It looked like a miniature solar system, the little moons racing around Jupiter as the planets were thought by Copernicus to orbit the Sun. It was another blow to the geocentrists. Forty-five years later, the celebrated Dutch physicist Christianus Huygens discovered a moon moving about the planet Saturn and named it Titan.* It was a dot of light a billion miles away, gleaming in reflected sunlight. From the time of its discovery, when European men wore long curly wigs, to World War II, when American men cut their hair down to stubble, almost nothing more was discovered about Titan except the fact it had a curious, tawny color. Ground-based telescopes could, even in principle, barely make out some enigmatic detail. The Spanish astronomer J. Comas Solá reported at the turn of the twentieth century some faint and indirect evidence of an atmosphere.

In a way, I grew up with Titan. I did my doctoral dissertation at the University of Chicago under the guidance of Gerard P. Kuiper, the astronomer who made the definitive discovery that Titan has an atmosphere. Kuiper was Dutch and in a direct line of intellectual descent from Christianus Huygens. In 1944, while making a spectroscopic examination of Titan, Kuiper was astonished to find the characteristic spectral features of the gas methane. When he pointed the telescope at Titan, there was the signature of methane.† When he pointed it away, not a hint of methane. But

* Not because he thought it remarkably large, but because in Greek mythology members of the generation preceding the Olympian gods—Saturn, his siblings, and his cousins—were called Titans.
† Titan's atmosphere has no detectable oxygen, so methane is not wildly out of chemical equilibrium—as it is on Earth—and its presence is in no way a sign of life.

moons were not supposed to hold onto sizable atmospheres, and the Earth's Moon certainly doesn't. Titan could retain an atmosphere, Kuiper realized, even though its gravity was less than Earth's, because its upper atmosphere is very cold. The molecules simply aren't moving fast enough for significant numbers to achieve escape velocity and trickle away to space.

Daniel Harris, a student of Kuiper's, showed definitively that Titan is red. Maybe we were looking at a rusty surface, like that of Mars. If you wanted to learn more about Titan, you could also measure the polarization of sunlight reflected off it. Ordinary sunlight is unpolarized. Joseph Veverka, now a fellow faculty member at Cornell University, was my graduate student at Harvard University, and therefore, so to speak, a grandstudent of Kuiper's. In *his* doctoral work, around 1970, he measured the polarization of Titan and found that it changed as the relative positions of Titan, the Sun, and the Earth changed. But the change was very different from that exhibited by, say, the Moon. Veverka concluded that the character of this variation was consistent with extensive clouds or haze on Titan. When we looked at it through the telescope, we weren't seeing its surface. We knew nothing about what the surface was like. We had no idea how far below the clouds the surface was.

So, by the early 1970s, as a kind of legacy from Huygens and his line of intellectual descent, we knew at least that Titan has a dense methane-rich atmosphere, and that it's probably enveloped by a reddish cloud veil or aerosol haze. But what kind of cloud is red? By the early 1970s my colleague Bishun Khare and I had been doing experiments at Cornell in which we irradiated various methane-rich atmospheres with ultraviolet light or electrons and were generating reddish or brownish solids; the stuff would coat the interiors of our reaction vessels. It seemed to me that, if methane-rich Titan had red-brown clouds, those clouds might very well be similar to what we were making in the laboratory. We called this material tholin, after a Greek word for "muddy." At the beginning we had very little idea what it was made of. It was some organic stew made by breaking apart our starting molecules, and allowing the atoms—carbon, hydrogen, nitrogen—and molecular fragments to recombine.

The simplest way to produce Titan tholin: Above, an electric current passes through a coil of copper wire wrapped around a glass tube, through which is flowing nitrogen and methane. The resulting electrical discharge breaks these gases down. The molecular fragments then recombine to form more complex materials. Below, after the experiment is over, the regions with most intense electrification during the experiment turn out to be most coated with the brownish solid we call Titan tholin. Similar results occur with more elaborate sources of energetic electrons. Experiments by Bishun N. Khare and the author, Laboratory for Planetary Studies, Cornell University.

The word "organic" carries no imputation of biological origin; following long-standing chemical usage dating back more than a century, it merely describes molecules built out of carbon atoms (excluding a few very simple ones such as carbon monoxide, CO, and carbon dioxide, CO_2). Since life on Earth is based on organic molecules, and since there was a time before there *was* life on Earth, some process must have made organic molecules on our planet before the time of the first organism. Something similar, I proposed, might be happening on Titan today.

The epochal event in our understanding of Titan was the arrival in 1980 and 1981 of the *Voyager 1* and *2* spacecraft in the Saturn system. The ultraviolet, infrared, and radio instruments revealed the pressure and temperature through the atmosphere—from the hidden surface to the edge of space. We learned how high the cloud tops are. We found that the air on Titan is composed mainly of nitrogen, N_2, as on the Earth today. The other principal constituent is, as Kuiper found, methane, CH_4, the starting material from which carbon-based organic molecules are generated there.

A variety of simple organic molecules was found, present as gases, mainly hydrocarbons and nitriles. The most complex of them have four "heavy" (carbon and/or nitrogen) atoms. Hydrocarbons are molecules composed of carbon and hydrogen atoms only, and are familiar to us as natural gas, petroleum, and waxes. (They're quite different from carbohydrates, such as sugars and starch, which also have oxygen atoms.) Nitriles are molecules with a carbon and nitrogen atom attached in a particular way. The best known nitrile is HCN, hydrogen cyanide, a deadly gas for humans. But hydrogen cyanide is implicated in the steps that on Earth led to the origin of life.

Finding these simple organic molecules in Titan's upper atmosphere—even if present only in a part per million or a part per billion—is tantalizing. Could the atmosphere of the primeval Earth have been similar? There's about ten times more air on Titan than there is on Earth today, but the early Earth may well have had a denser atmosphere.

Moreover, *Voyager* discovered an extensive region of energetic electrons and protons surrounding Saturn, trapped by the planet's magnetic field. During the course of its orbital motion around Saturn, Titan bobs in and out of this magnetosphere. Beams of electrons (plus solar ultraviolet light) fall on the upper air of Titan, just as charged particles (plus solar ultraviolet light) were intercepted by the atmosphere of the primitive Earth.

So it's a natural thought to irradiate the appropriate mixture of nitrogen and methane with ultraviolet light or electrons at very low pressures, and find out what more complex molecules can be made. Can we simulate what's going on in Titan's high atmosphere? In our laboratory at Cornell—with my colleague W. Reid Thompson playing a key role—we've replicated some of Titan's manufacture of organic gases. The simplest hydrocarbons on Titan are manufactured by ultraviolet light from the Sun. But for all the other gas products, those made most readily by electrons in the laboratory correspond to those discovered by *Voyager* on Titan, and in the same proportions. The correspondence is one to one. The next most abundant gases that we've found in the laboratory will be looked for in future studies of Titan. The most complex organic gases we make have six or seven carbon and/or nitrogen atoms. These product molecules are on their way to forming tholins.

WE HAD HOPED FOR A BREAK in the weather as *Voyager 1* approached Titan. A long distance away, it appeared as a tiny disk; at closest approach, our camera's field of view was filled by a small province of Titan. If there had been a break in the haze and clouds, even only a few miles across, as we scanned the disk we would have seen something of its hidden surface. But there was no hint of a break. This world is socked in. No one on Earth knows what's on Titan's surface. And an observer there, looking up in ordinary visible light, would have no idea of the glories that await upon ascending through the haze and beholding Saturn and its magnificent rings.

From measurements by *Voyager,* by the *International Ultraviolet Explorer* observatory in Earth orbit, and by ground-based telescopes, we know a fair amount about the orange-brown haze particles that obscure the surface: which colors of light they like to absorb, which colors they pretty much let pass through them, how

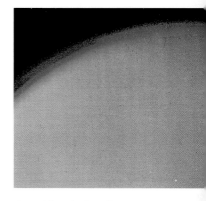

A world socked in: Titan as seen close up by *Voyager 1*. No breaks in the high-altitude haze could be found. The nature of the surface remains a mystery. Courtesy JPL/NASA.

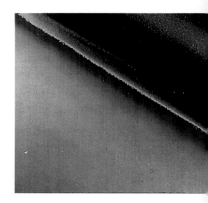

Detached haze layers (shown here in blue) above the main haze level on Titan. Their composition is unknown. Courtesy JPL/NASA.

The match of Titan tholin with the haze of Titan. In this diagram, *k* is a measure of how absorbing a given material is and is plotted on the vertical axis. The wavelength of light in microns (or micrometers) is plotted on the horizontal axis. Visible light is from 0.4 to 0.7 microns. Shorter than that are ultraviolet light and X-rays; longer than that are infrared and radio waves. Shaded brown are shown the properties of the Titan haze, as determined from ground-based optical telescopes, from the International Ultraviolet Explorer in Earth orbit, and from *Voyager* near Titan. The red line shows the values of *k* for the Titan tholin produced in our laboratory. The match is within the probable errors of measurement. Two other suggested organic solids, shown as black lines, the polymer of hydrogen cyanide and the polymer of acetylene, do not match the observations. From work by Bishun N. Khare and the author at Cornell University, and Edward Arakawa at Oak Ridge National Laboratory.

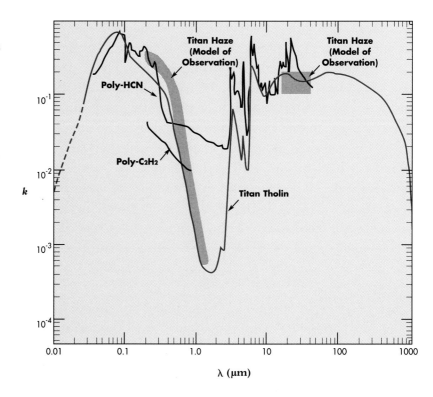

much they bend the light that does pass through them, and how big they are. (They're mostly the size of the particles in cigarette smoke.) The "optical properties" will depend, of course, on the composition of the haze particles.

In collaboration with Edward Arakawa of Oak Ridge National Laboratory in Tennessee, Khare and I have measured the optical properties of Titan tholin. It turns out to be a dead ringer for the real Titan haze. No other candidate material, mineral or organic, matches the optical constants of Titan. So we can fairly claim to have bottled the haze of Titan—formed high in its atmosphere, slowly falling out, and accumulating in copious amounts on its surface. What is this stuff made of?

It's very hard to know the exact composition of a complex organic solid. For example, the chemistry of coal is still not fully understood, despite a long-standing economic incentive. But we've found out some things about Titan tholin. It contains many of the essential building blocks of life on Earth. Indeed, if you drop Titan tholin into water you make a large number of amino acids, the fundamental constituents of proteins, and nucleotide bases also, the building blocks of DNA and RNA. Some of the amino acids so formed are widespread in living things on Earth. Others are of a completely different sort. A rich array of other organic molecules is present also, some relevant to life, some not. During the past four billion years, immense quantities of organic molecules sedimented out of the atmosphere onto the surface of Titan. If it's all deep-frozen and unchanged in the intervening aeons, the amount accumulated should be at least tens of meters (a hundred feet) thick; outside estimates put it at a kilometer deep.

But at 180°C below the freezing point of water, you might very well think that amino acids will never be made. Dropping tholins into water may be relevant to the early Earth, but not, it would seem, to Titan. However, comets and asteroids must on occasion come crashing into the surface of Titan. (The other nearby moons of Saturn show abundant impact craters, and the atmosphere of Titan isn't thick enough to prevent large, high-speed objects from reaching the surface.) Although we've never seen the surface of Titan, planetary scientists nevertheless know something about its composition. The average density of Titan lies between the density of ice and the density of rock. Plausibly it contains both. Ice and rock are abundant on nearby worlds, some of which are made of nearly pure ice. If the surface of Titan is icy, a high-speed cometary impact will temporarily melt the ice. Thompson and I estimate that any given spot on Titan's surface has a better than 50–50 chance of having once been melted, with an average lifetime of the impact melt and slurry of almost a thousand years.

This makes for a very different story. The origin of life on Earth seems to have occurred in oceans and shallow tidepools. Life on Earth is made mainly of water, which plays an essential physical and chemical role. Indeed, it's hard for us water-besotted creatures to imagine life without water. If on our planet the origin of life took less than a hundred million years, is there any chance that on Titan it took a thousand? With tholins mixed into liquid water—even for only a thousand years—the surface of Titan may be much further along towards the origin of life than we thought.

DESPITE ALL THIS, we understand pitifully little about Titan. This was brought home forcefully to me at a scientific symposium on Titan

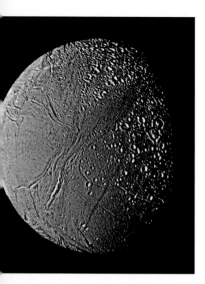

Another moon of Saturn: Enceladus, in false color (it's actually very white and ice-covered). A profusion of ancient impact craters can be seen to the right, but all at left have been eliminated by some melting event. *Voyager* image, courtesy JPL/NASA.

Another moon of Saturn: the trailing hemisphere of Dione. *Voyager* image, courtesy JPL/NASA.

held in Toulouse, France, and sponsored by the European Space Agency (ESA). While *oceans* of liquid water are impossible on Titan, oceans of liquid hydrocarbons are not. Clouds of methane (CH_4), the most abundant hydrocarbon, are expected not far above the surface. Ethane (C_2H_6), the next most abundant hydrocarbon, must condense out at the surface in the same way that water vapor becomes a liquid near the surface of the Earth, where the temperature is generally between the freezing and melting points. Vast oceans of liquid hydrocarbons should have accumulated over the lifetime of Titan. They would lie far beneath the haze and clouds. But that doesn't mean they would be wholly inaccessible to us—because radio waves readily penetrate the atmosphere of Titan and its suspended, slowly falling fine particles.

In Toulouse, Duane O. Muhleman of the California Institute of Technology described to us the very difficult technical feat of transmitting a set of radio pulses from a radio telescope in California's Mojave Desert, so they reach Titan, penetrate through the haze and clouds to its surface, are reflected back into space, and then returned to Earth. Here, the greatly enfeebled signal is picked up by an array of radio telescopes near Socorro, New Mexico. Great. If Titan has a rocky or icy surface, a radar pulse reflected off its surface should be detectable on Earth. But if Titan were covered with hydrocarbon oceans, Muhleman shouldn't see a thing: Liquid hydrocarbons are black to these radio waves, and no echo would have been returned to Earth. In fact, Muhleman's giant radar system sees a reflection when some longitudes of Titan are turned towards Earth, and not at other longitudes. All right, you might say, so Titan has oceans and continents, and it was a continent that reflected the signals back to Earth. But if Titan is in this respect like the Earth—for some meridians (through Europe and Africa, say) mainly continent, and for others (through the central Pacific, say) mainly ocean—then we must confront another problem:

The orbit of Titan around Saturn is not a perfect circle. It's noticeably squashed out, or elliptical. If Titan has extensive oceans, though, the giant planet Saturn around which it orbits will raise substantial tides on Titan, and the resulting tidal friction will circularize Titan's orbit in much less than the age of the Solar System.

In a 1982 scientific paper called "The Tide in the Seas of Titan," Stanley Dermott, now at the University of Florida, and I argued that for this reason Titan must be either an all-ocean or an all-land world. Otherwise the tidal friction in places where the ocean is shallow would have taken its toll. Lakes and islands might be permitted, but anything more and Titan would have a very different orbit than the one we see.

We have, then, three scientific arguments—one concluding that this world is almost entirely covered with hydrocarbon oceans, another that it's a mix of continents and oceans, and a third requiring us to choose, counseling that Titan can't have extensive oceans and extensive continents at the same time. It will be interesting to see what the answer turns out to be.

What I've just told you is a kind of scientific progress report. Tomorrow there might be a new finding that clears up these mysteries and contradictions. Maybe there's something wrong with Muhleman's radar results, although it's hard to see what it might be: His system tells him he's seeing Titan when it's nearest, when he ought to be seeing Titan. Maybe there's something wrong with

Voyager view of Iapetus. Negative image courtesy USGS/NASA.

Artist's conception of the surface of Iapetus, a bizarre world of ice and complex organic matter in an orbit exterior to Titan's. Saturn looms in its sky. Painting by Ron Miller.

The *Huygens* entry probe enters the haze of Titan on a parachute as its heat shield is separated. Painting courtesy Hamid Hassan, ESA.

Dermott's and my calculation about the tidal evolution of the orbit of Titan, but no one has been able to find any errors so far. And it's hard to see how ethane can avoid condensing out at the surface of Titan. Maybe, despite the low temperatures, over billions of years there's been a change in the chemistry; maybe some combination of comets impacting from the sky and volcanoes and other tectonic events, helped along by cosmic rays, can congeal liquid hydrocarbons, turning them into some complex organic solid that reflects radio waves back to space. Or maybe something reflective to radio waves is floating on the ocean surface. But liquid hydrocarbons are very underdense: Every known organic solid, unless extremely frothy, would sink like a stone in the seas of Titan.

Dermott and I now wonder whether, when we imagined continents and oceans on Titan, we were too transfixed by our experience on our own world, too Earth-chauvinist in our thinking.

The *Huygens* entry probe on the surface of Titan. The surface features are purely conjectural, since no one knows what will be found there. If we're lucky, the probe will continue to transmit data from the surface. Because of the haze, Saturn is not discernible in ordinary visible light from the surface of Titan, but perhaps this artist's conception shows the view in the near-infrared. Painting courtesy Hamid Hassan, ESA.

Another view of the *Huygens* entry probe about to touch down on the surface of Titan. The surface seems to be covered with dark, complex organic molecules. Painting courtesy Hamid Hassan, ESA.

Battered, cratered terrain and abundant impact basins cover other moons in the Saturn system. If we pictured liquid hydrocarbons slowly accumulating on one of those worlds, we would wind up not with global oceans, but with isolated large craters filled, although not to the brim, with liquid hydrocarbons. Many circular seas of petroleum, some over a hundred miles across, would be splattered across the surface—but no perceptible waves would be stimulated by distant Saturn and, it is conventional to think, no ships, no swimmers, no surfers, and no fishing. Tidal friction should, we calculate, be negligible in such a case, and Titan's stretched-out, elliptical orbit would not have become a so circular. We can't know for sure until we start getting radar or near-infrared images of the surface. But perhaps this is the resolution of our dilemma: Titan as a world of large circular hydrocarbon lakes, more of them in some longitudes than in others.

Should we expect an icy surface covered with deep tholin sediments, a hydrocarbon ocean with at most a few organic-encrusted islands poking up here and there, a world of crater lakes, or something more subtle that we haven't yet figured out? This isn't just an academic question, because there's a real spacecraft being designed to go to Titan. In a joint NASA/ESA program, a spacecraft called *Cassini* will be launched in October 1997—if all goes well. With two flybys of Venus, one of Earth, and one of Jupiter for gravitational assists, the ship will, after a seven-year voyage, be injected into orbit around Saturn. Each time the spacecraft comes close to Titan, the moon will be examined by an array of instruments, including radar. Because *Cassini* will be so much closer to Titan, it will be able to resolve many details on Titan's surface indetectable to Muhleman's pioneering Earth-based system. It's also likely that the surface can be viewed in the near infrared. Maps of the hidden surface of Titan may be in our hands sometime in the summer of 2004.

Cassini is also carrying an entry probe, fittingly called *Huygens,* which will detach itself from the main spacecraft and plummet into Titan's atmosphere. A great parachute will be deployed. The instrument package will slowly settle through the organic haze down into the lower atmosphere, through the methane clouds. It will examine organic chemistry as it descends, and—if it survives the landing—on the surface of this world as well.

Nothing is guaranteed. But the mission is technically feasible, hardware is being built, an impressive coterie of specialists, including many young European scientists, are hard at work on it, and all the nations responsible seem committed to the project. Perhaps it will actually come about. Perhaps winging across the billion miles of intervening interplanetary space will be, in the not too distant future, news about how far along the path to life Titan has come.

The *Cassini* spacecraft, foreground—on its way to a detailed reconaissance of Saturn, its rings, its magnetosphere, and its moons—launches the *Huygens* entry probe to Titan, lower left. Painting courtesy JPL/NASA.

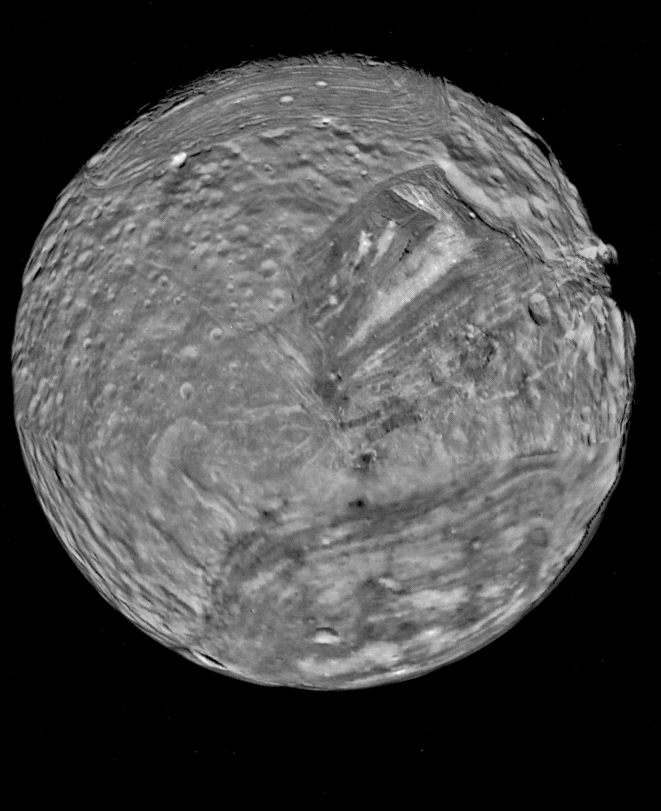

THE FIRST NEW PLANET

I implore you, you do not hope to be able to give the reasons
for the number of planets, do you?
This worry has been resolved . . .
—JOHANNES KEPLER,
EPITOME OF COPERNICAN ASTRONOMY,
BOOK 4 (1621)

B efore we invented civilization, our ancestors lived mainly in the open, out under the sky. Before we devised artificial lights and atmospheric pollution and modern forms of nocturnal entertainment, we watched the stars. There were practical calendrical reasons, of course, but there was more to it than that. Even today, the most jaded city dweller can be unexpectedly moved upon encountering a clear night sky studded with thou-

OPPOSITE: Uranus's satellite Miranda, perhaps the strangest moon in the Solar System. *Voyager* photomosaic, courtesy USGS/NASA.

sands of twinkling stars. When it happens to me, after all these years, it still takes my breath away.

In every culture, the sky and the religious impulse are intertwined. I lie back in an open field and the sky surrounds me. I'm overpowered by its scale. It's so vast and so far away that my own insignificance becomes palpable. But I don't feel rejected by the sky. I'm a part of it—tiny, to be sure, but everything is tiny compared to that overwhelming immensity. And when I concentrate on the stars, the planets, and their motions, I have an irresistible sense of machinery, clockwork, elegant precision working on a scale that, however lofty our aspirations, dwarfs and humbles us.

Most of the great inventions in human history—from stone tools and the domestication of fire to written language— were made by unknown benefactors. Our institutional memory of long-gone events is feeble. We do not know the name of that ancestor who first noted that planets were different from stars. She or he must have lived tens, perhaps even hundreds of thousands of years ago. But eventually people all over the world understood that five, no more, of the bright points of light that grace the night sky break lockstep with the others over a period of months, moving strangely—almost as if they had minds of their own.

Sharing the odd apparent motion of these planets were the Sun and Moon, making seven wandering bodies in all. These seven were important to the ancients, and they named them after gods—not any old gods, but the main gods, the chief gods, the ones who tell other gods (and mortals) what to do. One of the planets, bright and slow-moving, was named by the Babylonians after Marduk, by the Norse after Odin, by the Greeks after Zeus, and by the Romans after Jupiter, in each case the king of the gods. The faint, fast-moving one that was never far from the Sun the Romans named Mercury, after the messenger of the gods; the most brilliant of them was named Venus, after the goddess of love and beauty; the bloodred one Mars, after the god of war; and the most sluggish of the bunch Saturn, after the god of time. These metaphors and allusions were the best our ancestors could do: They possessed no scientific instruments beyond the naked eye, they were confined to the Earth, and they had no idea that it, too, is a planet.[*]

When it got to be time to design the week—a period of time, unlike the day, month, and year, with no intrinsic astronomical significance—it was assigned seven days, each named after one of the seven anomalous lights in the night sky. We can readily make out the remnants of this convention. In English, Saturday is Saturn's day. Sunday and Mo[o]nday are clear enough. Tuesday through Friday are named after the gods of the Saxon and kindred Teutonic invaders of Celtic/Roman Britain: Wednesday, for example, is Odin's (or Wodin's) day, which would be more apparent if we pronounced it as it's spelled, "Wedn's Day"; Thursday is Thor's day; Friday is the day of Freya, goddess of love. The last day of the week stayed Roman, the rest of it became German.

In all Romance languages, such as French, Spanish, and Italian, the connection is still more obvious, because they all derive from ancient Latin, in which the days of the week were named (in order, beginning with Sunday) after the Sun, the Moon, Mars, Mercury, Jupiter, Venus, and Saturn. (The Sun's day became the Lord's day.) They could have named the days in order of the brightness of the corresponding astronomical bodies—the Sun, the Moon, Venus, Jupiter, Mars, Saturn, Mercury (and thus Sunday,

---

[*] There was one moment in the last 4,000 years when all seven of these celestial bodies were clustered tightly together. Just before dawn on March 4, 1953 B.C., the crescent Moon was at the horizon. Venus, Mercury, Mars, Saturn, and Jupiter were strung out like jewels on a necklace near the great square in the constellation Pegasus—near the spot from which in our time the Perseid meteor shower emanates. Even casual watchers of the sky must have been transfixed by the event. What was it—a communion of the gods? According to the astronomer David Pankenier of Lehigh University and later Kevin Pang of JPL, this event was the starting point for the planetary cycles of the ancient Chinese astronomers.

There is no other time in the last 4,000 years (or in the next) when the dance of the planets around the Sun brings them so close together from the vantage point of Earth. But on May 5, 2000, all seven will be visible in the same part of the sky—although some at dawn and some at dusk and about ten times more spread out than on that late winter morning in 1953 B.C. Still, it's probably a good night for a party.

The lining up of Mercury, Mars, and Saturn (from left to right) with the Sun just emerging behind the Moon's night hemisphere. *Clementine* spacecraft image from lunar orbit. Courtesy Naval Research Laboratory.

Uranus with its five largest moons. *Voyager 2* photomontage from its January 1986 flyby. Right to left: Titania, Miranda, Oberon, Umbriel, and (left foreground) Ariel. Courtesy JPL/NASA.

Monday, Friday, Thursday, Tuesday, Saturday, Wednesday)—but they did not. If the days of the week in Romance languages had been ordered by distance from the Sun, the sequence would be Sunday, Wednesday, Friday, Monday, Tuesday, Thursday, Saturday. No one knew the order of the planets, though, back when we were naming planets, gods, and days of the week. The ordering of the days of the week seems arbitrary, although perhaps it does acknowledge the primacy of the Sun.

This collection of seven gods, seven days, and seven worlds—the Sun, the Moon, and the five wandering planets—entered the perceptions of people everywhere. The number seven began to acquire supernatural connotations. There were seven "heavens," the transparent spherical shells, centered on the Earth, that were imagined to make these worlds move. The outermost—the seventh heaven—is where the "fixed" stars were imagined to reside. There are Seven Days of Creation (if we include God's day of rest), seven orifices to the head, seven virtues, seven deadly sins, seven evil demons in Sumerian myth, seven vowels in the Greek alphabet (each affiliated with a planetary god), Seven Governors of Destiny according to the Hermetists, Seven Great Books of Manichaeism, Seven Sacraments, Seven Sages of Ancient Greece, and seven alchemical "bodies" (gold, silver, iron, mercury, lead, tin, and copper—gold still associated with the Sun, silver with the Moon, iron with Mars, etc.). The seventh son of a seventh son is endowed with supernatural powers. Seven is a "lucky" number. In the New Testament's Book of Revelations, seven seals on a scroll are opened, seven trumpets are sounded, seven bowls are filled. St. Augustine obscurely argued for the mystic importance of seven on the grounds that three "is the first whole number that is odd" (what about one?), "four the first that is even" (what about two?), and "of these . . . seven is composed." And so on. Even in our time these associations linger.

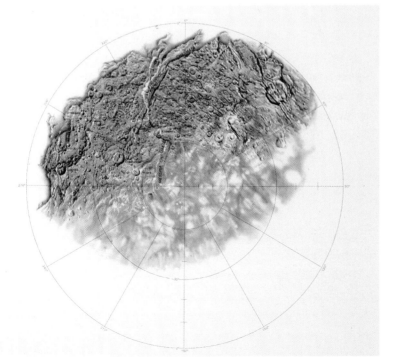

Titania, moon of Uranus, south polar projection. USGS shaded relief map.

Oberon, moon of Uranus, south polar projection. The craters on Oberon are named after heroes in Shakespearean plays. USGS shaded relief map.

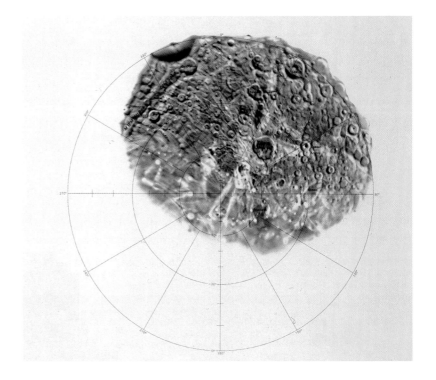

The existence even of the four satellites of Jupiter that Galileo discovered—hardly planets—was disbelieved on the grounds that it challenged the precedence of the number seven. As acceptance of the Copernican system grew, the Earth was added to the list of planets, and the Sun and Moon were removed. Thus, there seemed to be only six planets (Mercury, Venus, Earth, Mars, Jupiter, and Saturn). So learned academic arguments were invented showing why there *had* to be six. For example, six is the first "perfect" number, equal to the sum of its divisors (1 + 2 + 3). Q.E.D. And anyway, there were only six days of creation, not seven. People found ways to accommodate from seven planets to six.

As those adept at numerological mysticism adjusted to the Copernican system, this self-indulgent mode of thinking spilled over from planets to moons. The Earth had one moon; Jupiter had the four Galilean moons. That made five. Clearly one was missing. (Don't forget: Six is the first perfect number.) When Huygens discovered Titan in 1655, he and many others convinced themselves that it was the last: Six planets, six moons, and God's in His Heaven.

The historian of science I. Bernard Cohen of Harvard University has pointed out that Huygens actually gave up searching for other moons because it was apparent, from such arguments, that no more were to be found. Sixteen years later, ironically with Huygens in attendance, G. D. Cassini* of the Paris Observatory discovered a seventh moon—Iapetus, a bizarre world with one hemisphere black and the other white, in an orbit exterior to Titan's. Shortly after, Cassini discovered Rhea, the next Saturnian moon interior to Titan.

Here was another opportunity for numerology, this time harnessed to the practical task of flattering patrons. Cassini added up the number of planets (six) and the number of satellites (eight) and got fourteen. Now it so happened that the man who built Cassini's observatory for him and paid his salary was Louis XIV of France, the Sun King. The astronomer promptly "presented" these two new moons to his sovereign and proclaimed that Louis's "conquests" reached to the ends of the Solar System. Discreetly, Cassini then backed off from looking for more moons; Cohen suggests he was afraid one more might now offend Louis—a monarch not to be trifled with, who would shortly be throwing his subjects into dungeons for the crime of being Protestants. Twelve years later, though, Cassini returned to the search and found—doubtless with a measure of trepidation—another two moons. (It is probably a good thing that we have not continued in this vein; otherwise France would have been burdened by seventy-some-odd Bourbon kings named Louis.)

WHEN CLAIMS OF NEW WORLDS WERE MADE in the late eighteenth century, the force of such numerological arguments had much dissipated. Still, it was with a real sense of surprise that people heard in 1781 about a new planet, discovered through the telescope. New moons were comparatively unimpressive, especially after the first six or eight. But that there were new *planets* to be found and that humans had devised the means to do so were both considered astonishing, and properly so. If there is one previously unknown

---

* After whom the European-American mission to the Saturn system is named.

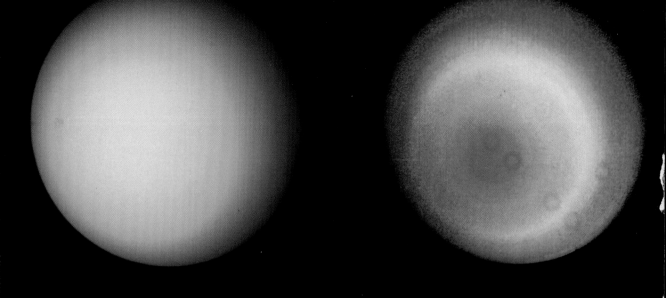

The bull's-eye in the sky: Uranus as imaged by *Voyager 2* on January 17, 1986 (at left as we would see the planet if we had been aboard the spacecraft; at right, in false color with greatly exaggerated contrasts and employing ultraviolet light). The south pole of the planet is pointed approximately at the Earth. The dark polar hood, seen here as orange, is probably composed of hydrocarbons generated by the electrons in Uranus's magnetic field as they pour into the polar atmosphere. The multicolored small circles in the right-hand image are caused by individual particles of dust in the *Voyager* camera optics. Courtesy JPL/NASA.

planet, there may be many more—in this solar system and in others. Who can tell what might be found if a multitude of new worlds are hiding in the dark?

The discovery was made not even by a professional astronomer but by William Herschel, a musician whose relatives had come to Britain with the family of another anglified German, the reigning monarch and future oppressor of the American colonists, George III. It became Herschel's wish to call the planet George ("George's Star," actually), after *his* patron, but, providentially, the name didn't stick. (Astronomers seem to have been very busy buttering up kings.) Instead, the planet that Herschel found is called Uranus (an inexhaustible source of hilarity renewed in each generation of English-speaking nine-year-olds). It is named after the ancient sky god who, according to Greek myth, was Saturn's father and thus the grandfather of the Olympian gods.

We no longer consider the Sun and Moon to be planets, and—ignoring the comparatively insignificant asteroids and comets—count Uranus as the seventh planet in order from the Sun (Mercury, Venus, Earth, Mars, Jupiter, Saturn, Uranus, Neptune, Pluto). It is the first planet unknown to the ancients. The four outer, Jovian, planets turn out to be very different from the four inner, terrestrial, planets. Pluto is a separate case.

As the years passed and the quality of astronomical instruments improved, we began to learn more about distant Uranus. What reflects the dim sunlight back to us is no solid surface, but atmosphere and clouds—just as for Titan, Venus, Jupiter, Saturn, and Neptune. The air on Uranus is made of hydrogen and helium, the two simplest gases. Methane and other hydrocarbons are also present. Just below the clouds visible to Earthbound observers is a massive atmosphere with enormous quantities of ammonia, hydrogen sulfide, and, especially, water.

At depth on Jupiter and Saturn, the pressures are so great that atoms sweat electrons, and the air becomes a metal. That does not seem to happen on less massive Uranus, because the pressures at depth are less. Still deeper, discovered only by its subtle tugs on Uranus' moons, wholly inaccessible to view, under the crushing weight of the overlying atmosphere, is a rocky surface. A big Earthlike planet is hiding down there, swathed in an immense blanket of air.

The Earth's surface temperature is due to the sunlight it intercepts. Turn off the Sun and the planet soon chills—not to trifling Antarctic cold, not just so cold that the oceans freeze, but to a cold so intense that the very air precipitates out, forming a ten-meter-thick layer of oxygen and nitrogen snows covering the whole planet. The little bit of energy that trickles up from the Earth's hot interior would be insufficient to melt these snows. For Jupiter, Saturn, and Neptune it's different. There's about as much heat pouring out from their interiors as they acquire from the warmth of the distant Sun. Turn off the Sun, and they would be only a little affected.

But Uranus is another story. Uranus is an anomaly among the Jovian planets. Uranus is like the Earth: There's very little intrinsic heat pouring out. We have no good understanding of why this should be, why Uranus—which in many respects is so similar to Neptune—should lack a potent source of internal heat. For this reason, among others, we cannot say we understand what is going on in the deep interiors of these mighty worlds.

Uranus is lying on its side as it goes around the Sun. In the 1990s, the south pole is heated by the Sun, and it is this pole that

Ariel, a moon of Uranus, south
polar projection. USGS shaded
relief map.

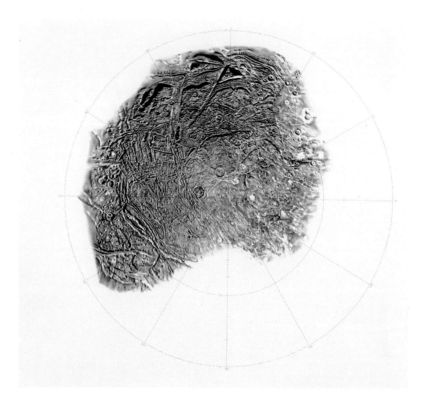

Earthbound observers at the end of the twentieth century see
when they look at Uranus. It takes Uranus 84 Earth years to make
one circuit of the Sun. So in the 2030s, the north pole will be sun-
ward (and Earthward). In the 2070s the south pole will be pointing to the Sun once again. In between, Earthbound astronomers
will be looking mainly at equatorial latitudes.

All the other planets spin much more upright in their orbits.
No one is sure of the reason for Uranus' anomalous spin; the most
promising suggestion is that sometime in its early history, billions
of years ago, it was struck by a rogue planet, about the size of the
Earth, in a highly eccentric orbit. Such a collision, if it ever happened, must have worked much tumult in the Uranus system; for
all we know, there may be other vestiges of ancient havoc still left
for us to find. But Uranus' remoteness tends to guard its mysteries.

In 1977 a team of scientists led by James Elliot, then of Cornell University, accidentally discovered that, like Saturn, Uranus

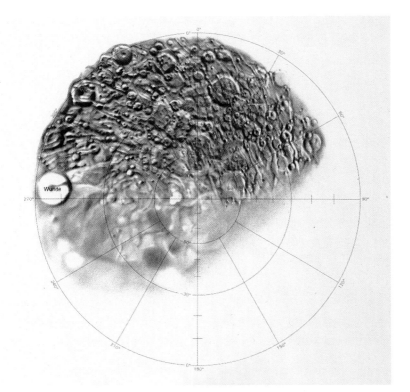

Umbriel, a moon of Uranus, south polar projection. USGS shaded relief map.

has rings. The scientists were flying over the Indian Ocean in a special NASA airplane—the Kuiper Airborne Observatory—to witness the passage of Uranus in front of a star. (Such passages, or occultations as they're called, happen from time to time, precisely because Uranus slowly moves with respect to the distant stars.) The observers were surprised to find that the star winked on and off several times just before it passed behind Uranus and its atmosphere, then several times more just after it emerged. Since the patterns of winking on and off were the same before and after occultation, this finding (and much subsequent work) has led to the discovery of nine very thin, very dark circumplanetary rings, giving Uranus the appearance of a bull's-eye in the sky.

Surrounding the rings, Earthbound observers understood, were the concentric orbits of the five moons then known: Miranda, Ariel, Umbriel, Titania, and Oberon. They're named after characters in Shakespeare's *A Midsummer Night's Dream* and *The*

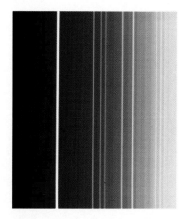

Close-up of the rings of Uranus in false color. The nine brightest lines (in groups, counting from the right, of three, two, three, and one) correspond to the nine known rings. The background pastel-colored lines are artifacts of the computer enhancement. The brightest or epsilon ring at left is neutral in color, whereas the eight other rings show real color differences, although greatly exaggerated in this image. Unlike the rings of Saturn, the rings of Uranus are very dark and thought to be made of radiation-processed organic matter. Courtesy JPL/NASA.

Tempest, and in Alexander Pope's *The Rape of the Lock.* Two of them were found by Herschel himself. The innermost of the five, Miranda, was discovered as recently as 1948, by my teacher G. P. Kuiper.* I remember how great an achievement the discovery of a new moon of Uranus was considered back then. The near-infrared light reflected by all five moons subsequently revealed the spectral signature of ordinary water ice on their surfaces. And no wonder—Uranus is so far from the Sun that it is no brighter there at noontime than it is after sunset on Earth. The temperatures are frigid. Any water must be frozen.

A REVOLUTION IN OUR UNDERSTANDING of the Uranus system—the planet, its rings, and its moons—began on January 24, 1986. On that day, after a journey of 8½ years, the *Voyager 2* spacecraft sailed very near to Miranda, and hit the bull's-eye in the sky. Uranus' gravity then flung it on to Neptune. The spacecraft returned 4,300 close-up pictures of the Uranus system and a wealth of other data.

Uranus was found to be surrounded by an intense radiation belt, electrons and protons trapped by the planet's magnetic field. *Voyager* flew through this radiation belt, measuring the magnetic field and the trapped charged particles as it went. It also detected—in changing timbres, harmonies, and nuance, but mainly in *fortissimo*—a cacaphony of radio waves generated by the speeding, trapped particles. Something similar was discovered on Jupiter and Saturn and would be later found at Neptune—but always with a theme and counterpoint characteristic of each world.

On Earth the magnetic and geographical poles are quite close together. On Uranus the magnetic axis and the axis of rotation are tilted away from each other by some 60 degrees. No one yet understands why: Some have suggested that we are catching Uranus in a reversal of its north and south magnetic poles, as periodically happens on Earth. Others propose that this too is the consequence

---

* He so named it because of the words spoken by Miranda, the heroine of *The Tempest:* "O brave new world, That has such people in't." (To which Prospero replies, "'Tis new to thee." Just so. Like all the other worlds in the Solar System, Miranda is about 4.5 billion years old.)

of that mighty, ancient collision that knocked the planet over. But we do not know.

Uranus is emitting much more ultraviolet light than it's receiving from the Sun, probably generated by charged particles leaking out of the magnetosphere and striking its upper atmosphere. From a vantage point in the Uranus system, the spacecraft examined a bright star winking on and off as the rings of Uranus passed by. New faint dust bands were found. From the perspective of Earth, the spacecraft passed behind Uranus; so the radio signals it was transmitting back home passed tangentially through the Uranian atmosphere, probing it—to below its methane clouds. A vast and deep ocean, perhaps 8,000 kilometers thick, of superheated liquid water floating in the air is inferred by some.

Among the principal glories of the Uranus encounter were the pictures. With *Voyager*'s two television cameras, we discovered ten new moons, determined the length of the day in the clouds of Uranus (about 17 hours), and studied about a dozen rings. The most spectacular pictures were those returned from the five larger, previously known moons of Uranus, especially the smallest of them, Kuiper's Miranda. Its surface is a tumult of fault valleys, parallel ridges, sheer cliffs, low mountains, impact craters, and frozen floods of once-molten surface material. This turmoiled landscape is unexpected for a small, cold, icy world so distant from the Sun. Perhaps the surface was melted and reworked in some long-gone epoch when a gravitational resonance between Uranus, Miranda, and Ariel pumped energy from the nearby planet into Miranda's interior. Or perhaps we are seeing the results of the primordial collision that is thought to have knocked Uranus over. Or, just conceivably, maybe Miranda was once utterly destroyed, dismembered, blasted into smithereens by a wild careening world, with many collision fragments still left in Miranda's orbit. The shards and remnants, slowly colliding, gravitationally attracting one another, may have reaggregated into just such a jumbled, patchy, unfinished world as Miranda is today.

For me, there's something eerie about the pictures of dusky Miranda, because I can remember so well when it was only a faint point of light almost lost in the glare of Uranus, discovered

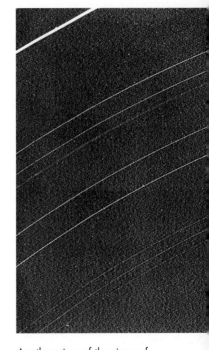

Another view of the rings of Uranus. The widest, the epsilon ring at top, is less than 100 kilometers (60 miles) wide. Some of the others are little more than 10 kilometers wide. Courtesy JPL/NASA.

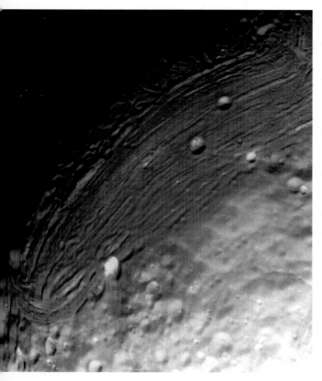
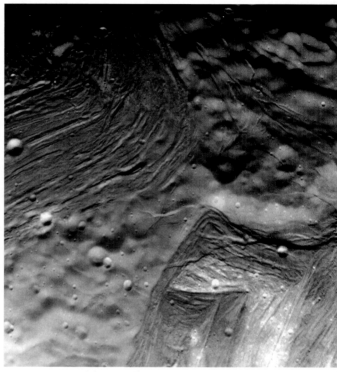

Two close-up views of Miranda obtained by *Voyager 2* on January 24, 1986. Whatever the origin of this bizarre terrain, the craters laid down on it indicate it is ancient, perhaps dating back to the formation of the Uranus system. Courtesy JPL/NASA.

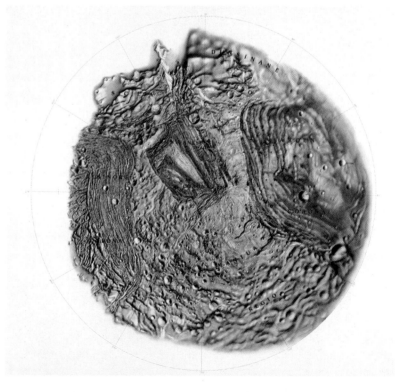

Miranda, a moon of Uranus, south polar projection. The nearly discontinuous boundaries between different parts of this world are real. USGS shaded relief map.

through great difficulty by dint of the astronomer's skills and patience. In only half a lifetime it has gone from an undiscovered world to a destination whose ancient and idiosyncratic secrets have been at least partially revealed.

Crescent Uranus as *Voyager 2* looks back. Courtesy USGS/NASA.

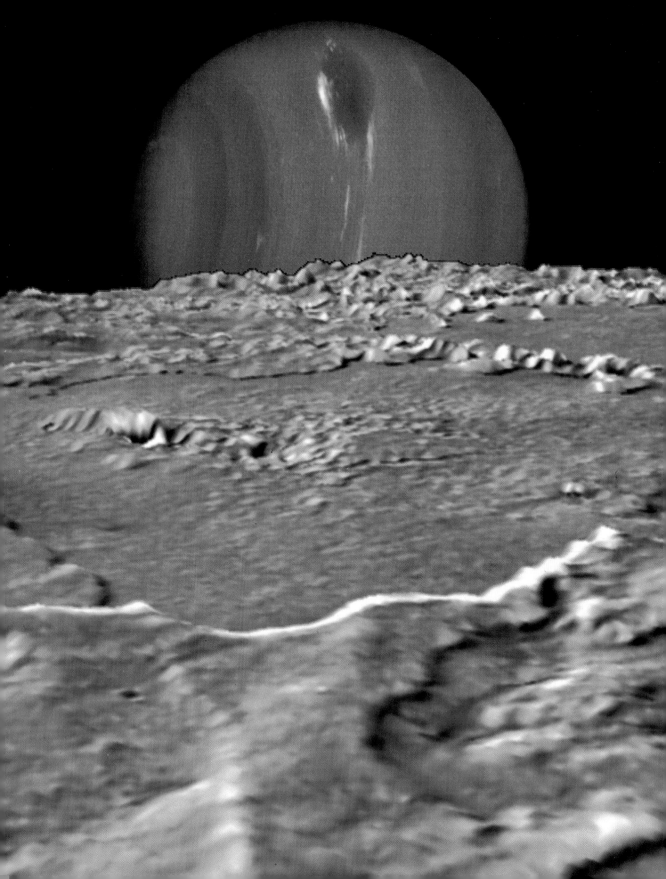

AN AMERICAN SHIP AT THE FRONTIERS OF THE SOLAR SYSTEM

... by the shore
Of Triton's Lake ...
I will clear my breast of secrets.
—EURIPIDES, *ION* (CA. 413 B.C.)

Neptune was the final port of call in *Voyager 2*'s grand tour of the Solar System. Usually, it is thought of as the penultimate planet, with Pluto the outermost. But because of Pluto's stretched-out, elliptical orbit, Neptune has lately been the outermost planet, and will remain so until 1999. Typical temperatures in its upper clouds are about -240°C, because it is so far from the warming rays of the Sun. It would be colder still, except for the heat welling up from its interior. Neptune glides along the hem of interstellar night. It is so far away that, in its sky, the Sun appears as little more than an extremely bright star.

OPPOSITE: Neptune as seen from just above the surface of its moon Triton. The cloud features in Neptune's atmosphere are moving, in this picture from top to bottom. *Voyager* photomontage, courtesy USGS/NASA.

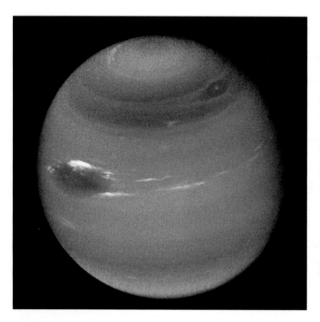 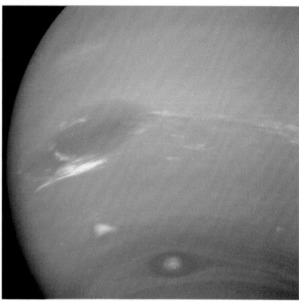

Neptune close-up. The three major cloud features, top to bottom at right, have been nicknamed "the Great Dark Spot," "the Scooter," and (with a bright core) "Dark Spot 2." They each rotate at different speeds, which is why the Scooter is present in the image at right but not in the one at left. They are all moving from west to east. We're looking at the clouds at the top of a deep atmosphere. Far below is a rocky core. Courtesy JPL/NASA.

How far? So far away that it has yet to complete a single trip around the Sun, a Neptunian year, since its discovery in 1846.[*] It's so far away that it cannot be seen with the naked eye. It's so far away that it takes light—faster than which nothing can go—more than five hours to get from Neptune to Earth.

When *Voyager 2* raced through the Neptune system in 1989, its cameras, spectrometers, particle and field detectors, and other instruments were feverishly examining the planet, its moons, and its rings. The planet itself, like its cousins Jupiter, Saturn, and Uranus, is a giant. Every planet is an Earthlike world at heart—but the four gas giants wear elaborate, cumbersome disguises. Jupiter and Saturn are great gas worlds with relatively small rocky and icy cores. But Uranus and Neptune are fundamentally rock and ice worlds swaddled in dense atmospheres that hide them from view.

Neptune is four times bigger than the Earth. When we look down on its cool, austere blueness, again we are seeing only atmosphere and clouds—no solid surface. Again, the atmosphere is made

[*] It takes so long to circuit the Sun because its orbit is so vast, 23 billion miles around, and because the force of the Sun's gravity—which keeps it from flying out into interstellar space—is at that distance comparatively feeble, less than a thousandth what it is in the Earth's vicinity.

mainly of hydrogen and helium, with a little methane and traces of other hydrocarbons. There may also be some nitrogen. The bright clouds, which seem to be methane crystals, float above thick, deeper clouds of unknown composition. From the motion of the clouds we discovered fierce winds, approaching the local speed of sound. A Great Dark Spot was found, curiously at almost the same latitude as the Great Red Spot on Jupiter. The azure color seems appropriate for a planet named after the god of the sea.

Surrounding this dimly lit, chilly, stormy, remote world is—here also—a system of rings, each composed of innumerable orbiting objects ranging in size from the fine particles in cigarette smoke to small trucks. Like the rings of the other Jovian planets, those of Neptune seem to be evanescent—it is calculated that gravity and solar radiation will disrupt them in much less than the age of the Solar System. If they are destroyed quickly, we must see them only because they were made recently. But how can rings be made?

The biggest moon in the Neptune system is called Triton.* Nearly six of our days are required for it to orbit Neptune, which —alone among big moons in the Solar System—it does in the opposite direction to which its planet spins (clockwise if we say Neptune rotates counterclockwise). Triton has a nitrogen-rich atmosphere, somewhat similar to Titan's; but, because the air and haze are much thinner, we can see its surface. The landscapes are varied and splendid. This is a world of ices—nitrogen ice, methane ice, probably underlain by more familiar water ice and rock. There are impact basins, which seem to have been flooded with liquid before refreezing (so there once were lakes on Triton); impact craters; long crisscrossing valleys; vast plains covered by freshly fallen nitrogen snow; puckered terrain that resembles the skin of a cantaloupe; and more or less parallel, long, dark streaks that seem

Neptune in south polar projection. Courtesy JPL/NASA.

Triton as seen by *Voyager 2* on its approach, some 4 million kilometers (2.5 million miles) away. Courtesy JPL/NASA.

---

* Robert Goddard, the inventor of the modern liquid-fueled rocket, envisioned a time when expeditions to the stars would be outfitted on and launched from Triton. This was in a 1927 afterthought to a 1918 handwritten manuscript called "The Last Migration." Considered much too daring for publication, it was deposited in a friend's safe. The cover page bears a warning: "The[se] notes should be read thoroughly only by an optimist."

Triton as revealed by *Voyager*. This picture will repay careful inspection. Where impact craters are rare, the surface, like that of Earth, must be young—that is, the craters have been filled in or covered over by some process. On this world, the process is thought to be oceans of methane or nitrogen that refroze, plus the seasonal coverage by methane and nitrogen snows. At top, note the profusion of dark streaks, all blown by the winds from west to east. There is much in this picture that is not well understood. *Voyager* photomosaic, courtesy USGS/NASA.

to have been blown by the wind, and then deposited on the icy surface—despite how sparse Triton's atmosphere is (about 1/10,000 the thickness of the Earth's).

All the craters on Triton are pristine—as if stamped out by some vast milling device. There are no slumped walls or muted relief. Even with the periodic falling and evaporation of snow, it seems that nothing has eroded the surface of Triton in billions of years. So the craters that were gouged out during the formation of Triton must have all been filled in and covered over by some early global resurfacing event. Triton orbits Neptune in the opposite direction to Neptune's rotation—unlike the situation with the Earth and its moon, and with most of the large moons in the Solar System. If Triton had formed out of the same spinning disk that made Neptune, it ought to be going around Neptune in the same direction that Neptune rotates. So Triton was not made from the original local nebula around Neptune, but arose somewhere else—perhaps far beyond Pluto—and was by chance gravitationally captured when it passed too close to Neptune. This event should have raised enormous solid-body tides in Triton, melting the surface and sweeping away all the past topography.

In some places the surface is as bright and white as freshly fallen Antarctic snows (and may offer a skiing experience unrivaled in all the Solar System). Elsewhere there's a tint, ranging from pink to brown. One possible explanation: Freshly fallen snows of nitrogen, methane, and other hydrocarbons are irradiated by solar ultraviolet light and by electrons trapped in the magnetic field of Neptune, through which Triton plows. We know that such irradiation will convert the snows (like the corresponding gases) to complex, dark, reddish organic sediments, ice tholins—nothing alive, but here too composed of some of the molecules implicated in the origin of life on Earth four billion years ago.

In local winter, layers of ice and snow build up on the surface. (Our winters, mercifully, are only 4 percent as long.) Through the spring, they are slowly transformed, more and more reddish organic molecules accumulating. By summertime, the ice and snow have evaporated; the gases so released migrate halfway across the

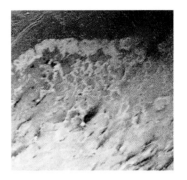

Close-up of some of the plumes of Triton. Courtesy JPL/NASA.

Artist's impression of a Triton plume. Breaking through to the surface after considerable pressures have been built up, a geyser of dark organics vents into the thin Tritonian atmosphere. In the background we see another plume caught by the prevailing winds. Painting by Ron Miller.

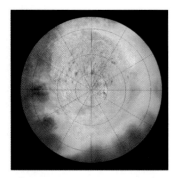

Triton as seen from just above its pole. Color differences have been greatly exaggerated. Courtesy USGS/NASA.

Neptune and its faint rings in the sky of Triton. Painting by Don Davis.

planet to the winter hemisphere and there cover the surface with ice and snow again. But the reddish organic molecules do not vaporize and are not transported—a lag deposit, they are next winter covered over by new snows, which are in turn irradiated, and by the following summer the accumulation is thicker. As time goes on, substantial amounts of organic matter are built up on the surface of Triton, which may account for its delicate color markings.

The streaks begin in small, dark source regions, perhaps when the warmth of spring and summer heats subsurface volatile snows. As they vaporize, gas comes gushing out as in a geyser, blowing off less-volatile surface snows and dark organics. Prevailing low-speed winds carry away the dark organics, which slowly sediment out of the thin air, are deposited on the ground, and generate the appearance of the streaks. This, at least, is one reconstruction of recent Tritonian history.

Triton may have large, seasonal polar caps of smooth nitrogen ice underlying layers of dark organic materials. Nitrogen snows

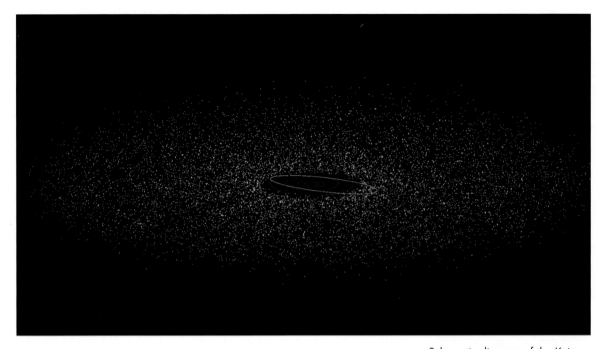

seem recently to have fallen at the equator. Snowfalls, geysers, windblown organic dust, and high-altitude hazes were entirely unexpected on a world with so thin an atmosphere.

Why is the air so thin? Because Triton is so far from the Sun. Were you somehow to pick this world up and move it into orbit around Saturn, the nitrogen and methane ices would quickly evaporate, a much denser atmosphere of gaseous nitrogen and methane would form, and radiation would generate an opaque tholin haze. It would become a world very like Titan. Conversely, if you moved Titan into orbit about Neptune, almost all its atmosphere would freeze out as snows and ices, the tholin would fall out and not be replaced, the air would clear, and the surface would become visible in ordinary light. It would become a world very like Triton.

These two worlds are not identical. The interior of Titan seems to contain much more ice than that of Triton, and much less rock. Titan's diameter is almost twice that of Triton. Still, if placed at the same distance from the Sun they would look like sisters. Alan Stern of the Southwest Research Institute suggests that they are two members of a vast collection of small worlds rich in

Schematic diagram of the Kuiper Comet Belt: Millions of small, icy worlds are thought to orbit the Sun just beyond Neptune and Pluto. The orbits of Jupiter, Saturn, Uranus, and Neptune are shown in violet, the orbit of Pluto in green. Pluto's orbit is tilted with respect to the orbits of the other planets. (Bearing that in mind, you can see why Pluto is sometimes not the outermost planet.) Far beyond the planets, and also far beyond the Kuiper Belt, is an enormous spherical array of icy worlds orbiting the Sun called the Oort Comet Cloud. Diagram by Harold Levison, Southwest Research Institute.

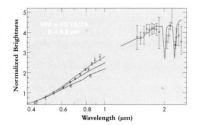

Pholus is one of the small, icy worlds recently found in the outermost provinces of the planetary part of our solar system. The Xs, squares, and other geometrical figures with attached error bars show the observed brightness of Pholus in different colors. From ultraviolet to near-infrared wavelengths, Pholus is brighter the longer the wavelength, meaning that it is very red. The reddest worlds in the Solar System are found near Uranus, Neptune, and beyond. The solid red line at top shows the best match yet achieved to the spectrum of Pholus: a mixture of complex organic solids, including tholins like those on Titan, plus ammonia ice. From work by Peter Wilson and the author, published in the February 1994 issue of *Icarus*.

nitrogen and methane that formed in the early Solar System. Pluto, yet to be visited by a spacecraft, appears to be another member of this group. Many more may await discovery beyond Pluto. The thin atmospheres and icy surfaces of all these worlds are being irradiated—by cosmic rays, if nothing else—and nitrogen-rich organic compounds are being formed. It looks as if the stuff of life is sitting not just on Titan, but throughout the cold, dimly lit outer reaches of our planetary system.

Another class of small objects has recently been discovered, whose orbits take them—at least part of the time—beyond Neptune and Pluto. Sometimes called minor planets or asteroids, they are more likely to be inactive comets (with no tails, of course; so far from the Sun, their ices cannot readily vaporize). But they are much bigger than the run-of-the-mill comets we know. They may be the vanguard of a vast array of small worlds that extends from the orbit of Pluto halfway to the nearest star. The innermost province of the Oort Comet Cloud, of which these new objects may be members, is called the Kuiper Belt, after my mentor Gerard Kuiper, who first suggested that it should exist. Short-period comets—like Halley's—arise in the Kuiper Belt, respond to gravitational tugs, sweep into the inner part of the Solar System, grow their tails, and grace our skies.

Back in the late nineteenth century, these building blocks of worlds—then mere hypotheses—were called "planetesimals." The flavor of the word is, I suppose, something like that of "infinitesimals": You need an infinite number of them to make anything. It's not quite *that* extreme with planetesimals, although a very large number of them would be required to make a planet. For example, trillions of bodies each a kilometer in size would be needed to coalesce to make a planet with the mass of the Earth. Once there were much larger numbers of worldlets in the planetary part of the Solar System. Most of them are now gone—ejected into interstellar space, fallen into the Sun, or sacrificed in the great enterprise of building moons and planets. But out beyond Neptune and Pluto the discards, the leftovers that were never aggregated into worlds, may be waiting—a few largish ones in the 100-kilometer range, and stupefying numbers of kilometer-sized and smaller bodies peppering the outer Solar System all the way out to the Oort Cloud.

In this sense there *are* planets beyond Neptune and Pluto—but they are not nearly as big as the Jovian planets, or even Pluto. Larger worlds may, for all we know, also be hiding in the dark beyond Pluto, worlds that can properly be called planets. The farther away they are, the less likely it is that we would have detected

Pluto and its moon Charon as photographed by the Hubble Space Telescope. This is the best photograph of both objects currently available. Pluto is redder than Charon. Pluto is smaller than the Earth's Moon and Charon is only about 1,270 kilometers (790 miles) in diameter. Courtesy R. Albrecht, ESA/ESO, and NASA.

A future spacecraft visits Pluto and its moon Charon. Surface features on these worlds are currently unknown, but the artist has for good reason imagined that Pluto is something like Triton. NASA artwork by Pat Rawlings/SAIC.

them. They cannot lie just beyond Neptune, though; their gravitational tugs would have perceptibly altered the orbits of Neptune and Pluto, and the *Pioneer 10* and *11* and *Voyager 1* and *2* spacecraft.

The newly discovered cometary bodies (with names like 1992QB and 1993FW) are not planets in this sense. If our detection threshold has just encompassed them, many more of them probably remain to be discovered in the outer Solar System—so far away that they're hard to see from Earth, so distant that it's a long journey to get to them. But small, quick ships to Pluto and beyond are within our ability. It would make good sense to dispatch one by Pluto and its moon Charon, and then, if we can, to make a close pass by one of the denizens of the Kuiper Comet Belt.

The rocky Earthlike cores of Uranus and Neptune seem to have accreted first, and then gravitationally attracted massive amounts of hydrogen and helium gas from the ancient nebula out of which the planets formed. Originally, they lived in a hailstorm. Their gravities were just sufficient to eject icy worldlets, when they came too close, far out beyond the realm of the planets, to populate the Oort Comet Cloud. Jupiter and Saturn became gas giants by the same process. But their gravities were too strong to populate the Oort Cloud: Ice worlds that came close to them were gravitationally pitched out of the Solar System entirely— destined to wander forever in the great dark between the stars.

So the lovely comets that on occasion rouse us humans to wonder and to awe, that crater the surfaces of inner planets and outer moons, and that now and then endanger life on Earth would be unknown and unthreatening had Uranus and Neptune not grown to be giant worlds four and a half billion years ago.

THIS IS THE PLACE for a brief interlude on planets *far* beyond Neptune and Pluto, planets of other stars.

Many nearby stars are surrounded by thin disks of orbiting gas and dust, often extending to hundreds of astronomical units (AU) from the local star (the outermost planets, Neptune and Pluto, are about 40 AU from our Sun). Younger Sun-like stars are

much more likely to have disks than older ones. In some cases, there's a hole in the center of the disk as in a phonograph record. The hole extends out from the star to perhaps 30 or 40 AU. This is true, for example, for the disks surrounding the stars Vega and Epsilon Eridani. The hole in the disk surrounding Beta Pictoris extends to only 15 AU from the star. There is a real possibility that these inner, dust-free zones have been cleaned up by planets that recently formed there. Indeed, this sweeping-out process is predicted for the early history of our planetary system. As observations improve, perhaps we will see telltale details in the configuration of dust and dust-free zones that will indicate the presence of planets too small and dark to be seen directly. Spectroscopic data suggest that these disks are churning and that matter is falling in on the central stars—perhaps from comets formed in the disk, deflected by the unseen planets, and evaporating as they approach too close to the local sun.

Because planets are small and shine by reflected light, they tend to be washed out in the glare of the local sun. Nevertheless, many efforts are now under way to find fully formed planets around nearby stars—by detecting a faint brief dimming of starlight as a dark planet interposes itself between the star and the observer on Earth; or by sensing a faint wobble in the motion of the star as it's tugged first one way and then another by an otherwise invisible orbiting companion. Spaceborne techniques will be much more sensitive. A Jovian planet going around a nearby star is about a billion times fainter than its sun; nevertheless, a new generation of ground-based telescopes that can compensate for the twinkling in the Earth's atmosphere may soon be able to detect such planets in only a few hours' observing time. A terrestrial planet of a neighboring star is a hundred times fainter still; but it now seems that comparatively inexpensive spacecraft, above the Earth's atmosphere, might be able to detect other Earths. None of these searches has succeeded yet, but we are clearly on the verge of being able to detect at least Jupiter-sized planets around the nearest stars—if there are any to be found.

A most important and serendipitous recent discovery is of a bona fide planetary system around an unlikely star, some 1,300 light-years away, found by a most unexpected technique: The pulsar designated B1257+12 is a rapidly rotating neutron star, an unbelievably dense sun, the remnant of a massive star that suffered a supernova explosion. It spins, at a rate measured to impressive precision, once every 0.0062185319388187 seconds. This pulsar is pushing 10,000 rpm.

Charged particles trapped in its intense magnetic field generate radio waves that are cast across the Earth, about 160 flickers a second. Small but discernible changes in the flash rate were tentatively interpreted by Alexander Wolszczan, now at Pennsylvania State University, in 1991—as a tiny reflex motion of the pulsar in response to the presence of planets. In 1994 the predicted mutual gravitational interactions of these planets were confirmed by Wolszczan from a study of timing residuals at the microsecond level over the intervening years. The evidence that these are truly new planets and not starquakes on the neutron star surface (or something) is now overwhelming—or, as Wolszczan put it, "irrefutable"; a new solar system is "unambiguously identified." Unlike all the other techniques, the pulsar timing method makes close-in terrestrial planets comparatively easy and more distant Jovian planets comparatively difficult to detect.

Planet C, some 2.8 times more massive than the Earth, orbits

The Vela supernova remnant. Could planets survive a supernova explosion? The line is a rotating artificial satellite of the Earth captured as it crosses the telescope's field of view in this time exposure. Courtesy Anglo-Australian Observatory. Photograph by David Malin.

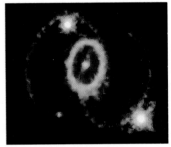

Three rings of glowing gas surrounding the site of the supernova called 1987A. It lies in the Large Magellanic Cloud, a dwarf satellite galaxy of the Milky Way about 169,000 light-years from Earth. Astronomers on Earth observed the explosion in February 1987, but of course it occurred 169,000 years ago. Is it possible that new planets can form from such rings, the remnants of supernova explosions? Hubble Space Telescope picture, courtesy Christopher Burrows, European Space Agency/Space Telescope Science Institute, and NASA.

the pulsar every 98 days at a distance of 0.47 astronomical units [*] (AU); Planet B, with about 3.4 Earth masses, has a 67–Earth-day year at 0.36 AU. A smaller world, Planet A, still closer to the star, with about 0.015 Earth masses, is at 0.19 AU. Crudely speaking, Planet B is roughly at the distance of Mercury from our Sun; Planet C is midway between the distances of Mercury and Venus; and interior to both of them is Planet A, roughly the mass of the

[*] The Earth, by definition, is 1 AU from its star, the Sun.

Moon at about half Mercury's distance from our Sun. Whether these planets are the remnants of an earlier planetary system that somehow survived the supernova explosion that produced the pulsar, or whether they formed from the resulting circumstellar accretion disk subsequent to the supernova explosion, we do not know. But in either case, we have now learned that there are other Earths.

The energy put out by B1257+12 is about 4.7 times that of the Sun. But, unlike the Sun, most of this is not in visible light, but in a fierce hurricane of electrically charged particles. Suppose that these particles impinge on the planets and heat them. Then, even a planet at 1 AU would have a surface around 280 Celsius degrees above the normal boiling point of water, greater than the temperature of Venus.

These dark and broiling planets do not seem hospitable for life. But there may be others, farther from B1257+12, that are. (Hints of at least one cooler, outer world in the B1257+12 system exist.) Of course, we don't even know that such worlds would retain their atmospheres; perhaps any atmospheres were stripped away in the supernova explosion, if they date back that far. But we do seem to be detecting a recognizable planetary system. Many more are likely to become known in coming decades, around ordinary Sun-like stars as well as white dwarfs, pulsars, and other end states of stellar evolution.

Eventually, we will have a list of planetary systems—each perhaps with terrestrials and Jovians and maybe new classes of planets. We will examine these worlds, spectroscopically and in other ways. We will be searching for new Earths and other life.

ON NONE OF THE WORLDS in the outer Solar System did *Voyager* find signs of life, much less intelligence. There was organic matter galore—the stuff of life, the premonitions of life, perhaps—but as far as we could see, no life. There was no oxygen in their atmospheres, and no gases profoundly out of chemical equilibrium, as methane is in the Earth's oxygen. Many of the worlds were painted with subtle colors, but none with such distinctive, sharp absorption features as chlorophyll provides over much of the Earth's surface. On very few worlds was *Voyager* able to resolve details as small as a kilometer across. By this standard, it would not have detected even our own technical civilization had it been transplanted to the outer Solar System. But for what it's worth, we found no regular patterning, no geometrization, no passion for small circles, triangles, squares, or rectangles. There were no constellations of steady points of light on the night hemispheres. There were no signs of a technical civilization reworking the surface of any of these worlds.

The Jovian planets are prolific broadcasters of radio waves— generated in part by the abundant trapped and beamed charged particles in their magnetic fields, in part by lightning, and in part by their hot interiors. But none of this emission has the character of intelligent life—or so it seems to the experts in the field.

Of course our thinking may be too narrow. We may be missing something. For example, there is a little carbon dioxide in the atmosphere of Titan, which puts its nitrogen/methane atmosphere out of chemical equilibrium. I think the CO_2 is provided by the steady pitter-patter of comets falling into Titan's atmosphere—but maybe not. Maybe there's something on the surface unaccountably generating CO_2 in the face of all that methane.

The surfaces of Miranda and Triton are unlike anything else we know. There are vast chevron-shaped landforms and crisscrossing straight lines that even sober planetary geologists once mischievously described as "highways." We think we (barely) understand these landforms in terms of faults and collisions, but of course we might be wrong.

The surface stains of organic matter—sometimes, as on Triton, delicately hued—are attributed to charged particles producing chemical reactions in simple hydrocarbon ices, generating more complex organic materials, and all this having nothing to do with the intermediation of life. But of course we might be wrong.

The complex pattern of radio static, bursts, and whistles that we receive from all four Jovian planets seems, in a general way, explicable by plasma physics and thermal emission. (Much of the detail is not yet well understood.) But of course we might be wrong.

We have found nothing on dozens of worlds so clear and

Diagram of the Solar System embedded in the solar wind: the Sun's extended atmosphere blowing out into interstellar space. Four operational spacecraft are racing out of the Solar System and have a chance of detecting the boundary between the wind from the Sun and the wind from the stars before their power fails: *Pioneers 10* and *11*, shown as red arrows, and *Voyagers 1* and *2*, shown as yellow arrows. The *Voyagers* are traveling faster and will retain transmitter power further into the future. From *EOS Transactions*, April 19, 1994, Courtesy American Geophysical Union.

striking as the signs of life found by the *Galileo* spacecraft in its passages by the Earth. Life is a hypothesis of last resort. You invoke it only when there's no other way to explain what you see. If I had to judge, I would say that there's no life on any of the worlds we've studied, except of course our own. But I might be wrong, and, right or wrong, my judgment is necessarily confined to this Solar System. Perhaps on some new mission we'll find something different, something striking, something wholly inexplicable with the ordinary tools of planetary science—and tremulously, cautiously, we will inch toward a biological explanation. However, for now nothing requires that we go down such a path. So far, the only life in the Solar System is that which comes from Earth. In the Uranus and Neptune systems, the only sign of life has been *Voyager* itself.

As we identify the planets of other stars, as we find other worlds of roughly the size and mass of the Earth, we will scrutinize *them* for life. A dense oxygen atmosphere may be detectable even on a world we've never imaged. As for the Earth, that may by itself

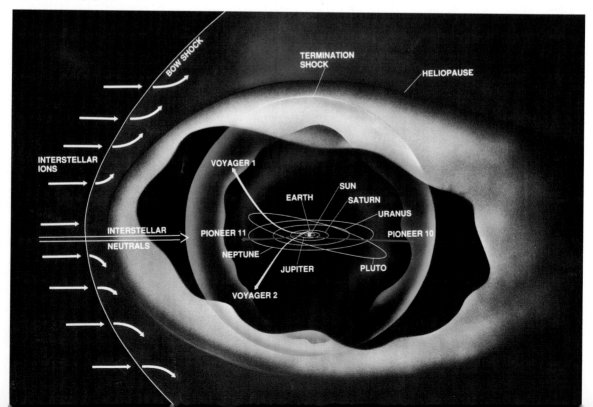

be a sign of life. An oxygen atmosphere with appreciable quantities of methane would almost certainly be a sign of life, as would modulated radio emission. Someday, from observations of our planetary system or another, the news of life elsewhere may be announced over the morning coffee.

THE *VOYAGER* SPACECRAFT are bound for the stars. They are on escape trajectories from the Solar System, barreling along at almost a million miles a day. The gravitational fields of Jupiter, Saturn, Uranus, and Neptune have flung them at such high speeds that they have broken the bonds that once tied them to the Sun.

Have they left the Solar System yet? The answer depends very much on how you define the boundary of the Sun's realm. If it's the orbit of the outermost good-sized planet, then the *Voyager* spacecraft are already long gone; there are probably no undiscovered Neptunes. If you mean the outermost planet, it may be that there are other—perhaps Triton-like—planets far beyond Neptune and Pluto; if so, *Voyager 1* and *Voyager 2* are still within the Solar System. If you define the outer limits of the Solar System as the heliopause—where the interplanetary particles and magnetic fields are replaced by their interstellar counterparts—then neither *Voyager* has yet left the Solar System, although they may do so in the next few decades.[*] But if your definition of the edge of the Solar System is the distance at which our star can no longer hold worlds in orbit about it, then the *Voyagers* will not leave the Solar System for hundreds of centuries.

Weakly grasped by the Sun's gravity, in every direction in the

[*] Radio signals that both *Voyagers* detected in 1992 are thought to arise from the collision of powerful gusts of solar wind with the thin gas that lies between the stars. From the immense power of the signal (over 10 trillion watts), the distance to the heliopause can be estimated: about 100 times the Earth's distance from the Sun. At the speed it's leaving the Solar System, *Voyager 1* might pierce the heliopause and enter interstellar space around the year 2010. If its radioactive power source is still working, news of the crossing will be radioed back to the stay-at-homes on Earth. The energy released by the collision of this shock wave with the heliopause makes it the most powerful source of radio emission in the Solar System. It makes you wonder whether even stronger shocks in other planetary systems might be detectable by our radio telescopes.

sky, is that immense horde of a trillion comets or more, the Oort Cloud. The two spacecraft will finish their passage through the Oort Cloud in another 20,000 years or so. Then, at last, completing their long good-bye to the Solar System, broken free of the gravitational shackles that once bound them to the Sun, the *Voyagers* will make for the open sea of interstellar space. Only then will Phase Two of their mission begin.

Their radio transmitters long dead, the spacecraft will wander for ages in the calm, cold interstellar blackness—where there is almost nothing to erode them. Once out of the Solar System, they will remain intact for a billion years or more, as they circumnavigate the center of the Milky Way galaxy.

We do not know whether there are other spacefaring civilizations in the Milky Way. If they do exist, we do not know how abundant they are, much less where they are. But there is at least a chance that sometime in the remote future one of the *Voyagers* will be intercepted and examined by an alien craft.

Accordingly, as each *Voyager* left Earth for the planets and the stars, it carried with it a golden phonograph record encased in a golden, mirrored jacket containing, among other things: greetings in 59 human languages and one whale language; a 12-minute sound essay including a kiss, a baby's cry, and an EEG record of the meditations of a young woman in love; 116 encoded pictures, on our science, our civilization, and ourselves; and 90 minutes of the Earth's greatest hits—Eastern and Western, classical and folk, including a Navajo night chant, a Japanese *shakuhachi* piece, a Pygmy girl's initiation song, a Peruvian wedding song, a 3,000-year-old composition for the *ch'in* called "Flowing Streams," Bach, Beethoven, Mozart, Stravinsky, Louis Armstrong, Blind Willie Johnson, and Chuck Berry's "Johnny B. Goode."

Space is nearly empty. There is virtually no chance that one of the *Voyagers* will ever enter another solar system—and this is true even if every star in the sky is accompanied by planets. The instructions on the record jackets, written in what we believe to be readily comprehensible scientific hieroglyphics, can be read, and the contents of the records understood, only if alien beings, somewhere in the remote future, find *Voyager* in the depths of interstellar space. Since both *Voyagers* will circle the center of the Milky Way Galaxy essentially forever, there is plenty of time for the records to be found—if there's anyone out there to do the finding.

We cannot know how much of the records they would understand. Surely the greetings will be incomprehensible, but their intent may not be. (We thought it would be impolite not to say hello.) The hypothetical aliens are bound to be very different from us—independently evolved on another world. Are we really sure they could understand anything at all of our message? Every time I feel these concerns stirring, though, I reassure myself: Whatever the incomprehensibilities of the *Voyager* record, any alien ship that finds it will have another standard by which to judge us. Each *Voyager* is itself a message. In their exploratory intent, in the lofty ambition of their objectives, in their utter lack of intent to do harm, and in the brilliance of their design and performance, these robots speak eloquently for us.

But being much more advanced scientists and engineers than we—otherwise they would never be able to find and retrieve the small, silent spacecraft in interstellar space—perhaps the aliens would have no difficulty understanding what is encoded on these golden records. Perhaps they would recognize the tentativeness of our society, the mismatch between our technology and our wisdom. Have we destroyed ourselves since launching *Voyager,* they might wonder, or have we gone on to greater things?

Or perhaps the records will never be intercepted. Perhaps no one in five billion years will ever come upon them. Five billion years is a long time. In five billion years, all humans will have become extinct or evolved into other beings, none of our artifacts will have survived on Earth, the continents will have become unrecognizably altered or destroyed, and the evolution of the Sun will have burned the Earth to a crisp or reduced it to a whirl of atoms.

Far from home, untouched by these remote events, the *Voyagers,* bearing the memories of a world that is no more, will fly on.

CHAPTER 1 0

SACRED BLACK

Deep sky is, of all visual impressions, the nearest akin to a feeling.
—SAMUEL TAYLOR COLERIDGE, *NOTEBOOKS* (1805)

The blue of a cloudless May morning, or the reds and oranges of a sunset at sea, have roused humans to wonder, to
poetry, and to science. No matter where on Earth we live,
no matter what our language, customs, or politics, we share a sky
in common. Most of us *expect* that azure blue and would, for good
reason, be stunned to wake up one sunrise to find a cloudless sky
that was black or yellow or green. (Inhabitants of Los Angeles and
Mexico City have grown accustomed to brown skies, and those of
London and Seattle to gray ones—but even they still consider blue
the planetary norm.)

And yet there *are* worlds with black or yellow skies, and
maybe even green. The color of the sky characterizes the world.
Plop me down on any planet in the Solar System; without sensing

OPPOSITE: The band of blue is the
Earth's atmosphere seen edge-on
from the space shuttle *Discovery*
on Mission 41-D off the coast of
Rio de Janeiro, Brazil. It is sunset.
Thunderheads can be seen
penetrating into the stratosphere.
Beyond the band of blue is the
blackness of space. Courtesy
Johnson Space Center/NASA.

Every photo of the horizon from low orbit shows the band of blue, as in this shuttle image of a tropical storm. Courtesy Johnson Space Center/NASA.

the gravity, without glimpsing the ground, let me take a quick look at the Sun and sky, and I can, I think, pretty well tell you where I am. That familiar shade of blue, interrupted here and there by fleecy white clouds, is a signature of our world. The French have an expression, *sacre-bleu!*, which translates roughly as "Good heavens!" [*] Literally, it means "sacred blue!" Indeed. If there ever is a true flag of Earth, this should be its color.

Birds fly through it, clouds are suspended in it, humans admire and routinely traverse it, light from the Sun and stars flutters through it. But what *is* it? What is it made of? Where does it end? How much of it is there? Where does all that blue come from? If it's a commonplace for all humans, if it typifies our world, surely we should know something about it. What is the sky?

In August 1957, for the first time, a human being rose above the blue and looked around—when David Simons, a retired Air Force officer and a physician, became the highest human in history. Alone, he piloted a balloon to an altitude of over 100,000 feet (30 kilometers) and through his thick windows glimpsed a

[*] Like "gosh-darned" and "geez," this phrase was originally a euphemism for those who considered *Sacre-Dieu!*, "Sacred God!," too strong an oath, the Second Commandment duly considered, to be uttered aloud.

different sky. Now a professor at the University of California Medical School in Irvine, Dr. Simons recalls it was a dark, deep purple overhead. He had reached the transition region where the blue of ground level is being overtaken by the perfect black of space.

Since Simons' almost forgotten flight, people of many nations have flown above the atmosphere. It is now clear from repeated and direct human (and robotic) experience that in space the daytime sky is black. The Sun shines brightly on your ship. The Earth below you is brilliantly illuminated. But the sky above is black as night.

Here is the memorable description by Yuri Gagarin of what he saw on the first spaceflight of the human species, aboard *Vostok 1,* on April 12, 1961:

> The sky is completely black; and against the background of this black sky the stars appear somewhat brighter and more distinct. The Earth has a very characteristic, very beautiful blue halo, which is seen well when you observe the horizon. There is a smooth color transition from tender blue, to blue, to dark blue

The Earth and the Moon in the blackness of space. As we rise above either world, the sky must become black. *Galileo* image, courtesy JPL/NASA.

and purple, and then to the completely black color of the sky. It is a very beautiful transition.

Clearly, the daylit sky—all that blue—is somehow connected with the air. But as you look across the breakfast table, your companion is not (usually) blue; the color of the sky must be a property not of a little air, but of a great deal. If you look closely at the Earth from space, you see it surrounded by a thin band of blue, as thick as the lower atmosphere; indeed, it *is* the lower atmosphere. At the top of that band you can make out the blue sky fading into the blackness of space. This is the transition zone that Simons was the first to enter and Gagarin the first to observe from above. In routine spaceflight, you start at the bottom of the blue, penetrate entirely through it a few minutes after liftoff, and then enter that boundless realm where a simple breath of air is impossible without elaborate life-support systems. Human life depends for its very existence on that blue sky. We are right to consider it tender and sacred.

We see the blue in daytime because sunlight is bouncing off the air around and above us. On a cloudless night, the sky is black because there is no sufficiently intense source of light to be reflected off the air. Somehow, the air preferentially bounces blue light down to us. How?

The visible light from the Sun comes in many colors—violet, blue, green, yellow, orange, red—corresponding to light of different wavelengths. (A wavelength is the distance from crest to crest as the wave travels through air or space.) Violet and blue light waves have the shortest wavelengths; orange and red the longest. What we perceive as color is how our eyes and brains read the wavelengths of light. (We might just as reasonably translate wavelengths of light into, say, heard tones rather than seen colors—but that's not how our senses evolved.)

When all those rainbow colors of the spectrum are mixed together, as in sunlight, they seem almost white. These waves travel together in eight minutes across the intervening 93 million miles (150 million kilometers) of space from the Sun to the Earth. They strike the atmosphere, which is made mostly of nitrogen and oxygen molecules. Some waves are reflected by the air back into space. Some are bounced around before the light reaches the ground and they can be detected by a passing eyeball. (Also, some bounce off clouds or the ground back into space.) This bouncing around of light waves in the atmosphere is called "scattering."

But not all waves are equally well scattered by the molecules of air. Wavelengths that are much longer than the size of the molecules are scattered less; they spill over the molecules, hardly influenced by their presence. Wavelengths that are closer to the size of the molecules are scattered more. And waves have trouble ignoring obstacles as big as they are. (You can see this in water waves scattered by the pilings of piers, or bathtub waves from a dripping faucet encountering a rubber duck.) The shorter wavelengths, those that we sense as violet and blue light, are more efficiently scattered than the longer wavelengths—those that we sense as orange and red light. When we look up on a cloudless day and admire the blue sky, we are witnessing the preferential scattering of the short waves in sunlight. This is called Rayleigh scattering, after the English physicist who offered the first coherent explanation for it. Cigarette smoke is blue for just the same reason: The particles that make it up are about as small as the wavelength of blue light.

So why is the sunset red? The red of the sunset is what's left of sunlight after the air scatters the blue away. Since the atmosphere is a thin shell of gravitationally bound gas surrounding the solid Earth, sunlight must pass through a longer slant path of air at sunset (or sunrise) than at noon. Since the violet and blue waves are scattered even more during their now-longer path through the air than when the Sun is overhead, what we see when we look toward the Sun is the residue—the waves of sunlight that are hardly scattered away at all, especially the oranges and reds. A blue sky makes a red sunset. (The noontime Sun seems yellowish partly because it emits slightly more yellow light than other colors, and partly because, even with the Sun overhead, some blue light is scattered out of the sunbeams by the Earth's atmosphere.)

It is sometimes said that scientists are unromantic, that their passion to figure out robs the world of beauty and mystery. But is it not stirring to understand how the world actually works—that white light is made of colors, that color is the way we perceive the wavelengths of light, that transparent air reflects light, that in so doing it discriminates among the waves, and that the sky is blue for the same reason that the sunset is red? It does no harm to the romance of the sunset to know a little bit about it.

Since most simple molecules are about the same size (roughly a hundred millionth of a centimeter), the blue of the Earth's sky doesn't much depend on what the air is made of—as long as the air doesn't *absorb* the light. Oxygen and nitrogen molecules don't absorb visible light; they only bounce it away in some other direction. Other molecules, though, can gobble up the light. Oxides of nitrogen—produced in automotive engines and in the fires of industry—are a source of the murky brown coloration of smog. Oxides of nitrogen (made from oxygen and nitrogen) *do* absorb light. Absorption, as well as scattering, can color a sky.

OTHER WORLDS, OTHER SKIES: Mercury, the Earth's Moon, and most satellites of the other planets are small worlds; because of their feeble gravities, they are unable to retain their atmospheres—which instead trickle off into space. The near-vacuum of space then reaches the ground. Sunlight strikes their surfaces unimpeded, neither scattered nor absorbed along the way. The skies of these worlds are black, even at noon. This has been witnessed firsthand so far by only 12 humans, the lunar landing crews of *Apollos 11, 12,* and *14–17.*

A full list of the satellites in the Solar System, known as of this writing, is given in the accompanying table. (Nearly half of them were discovered by *Voyager.*) All have black skies—except Titan of Saturn and perhaps Triton of Neptune, which are big enough to have atmospheres. And all asteroids as well.

Venus has about 90 times more air than Earth. It isn't mainly oxygen and nitrogen as here—it's carbon dioxide. But carbon dioxide doesn't absorb visible light either. What would the sky look like from the surface of Venus if Venus had no clouds? With so much atmosphere in the way, not only are violet and blue waves scattered, but all the other colors as well—green, yellow, orange, red. The air is so thick, though, that hardly any blue light makes it to the ground; it's scattered back to space by successive bounces higher up. Thus, the light that does reach the ground should be

SIXTY-TWO WORLDS FOR THE THIRD MILLENNIUM: KNOWN MOONS OF THE PLANETS (AND ONE ASTEROID)—LISTED IN ORDER OF DISTANCE FROM THEIR PLANET

EARTH, 1	MARS, 2	IDA, 1	JUPITER, 16	SATURN, 18	URANUS, 15	NEPTUNE, 8	PLUTO, 1
Moon	Phobos	Dactyl	Metis	Pan	Cordelia	Naiad	Charon
	Deimos		Adrastea	Atlas	Ophelia	Thalassa	
			Amalthea	Prometheus	Bianca	Despina	
			Thebe	Pandora	Cressida	Galatea	
			Io	Epimetheus	Desdemona	Larissa	
			Europa	Janus	Juliet	Proteus	
			Ganymede	Mimas	Portia	Triton	
			Callisto	Enceladus	Rosalind	Nereid	
			Leda	Tethys	Belinda		
			Himalia	Telesto	Puck		
			Lysithea	Calypso	Miranda		
			Elara	Dione	Ariel		
			Ananke	Helene	Umbriel		
			Carme	Rhea	Titania		
			Pasiphae	Titan	Oberon		
			Sinope	Hyperion			
				Iapetus			
				Phoebe			

strongly reddened—like an Earth sunset all over the sky. Further, sulfur in the high clouds will stain the sky yellow. Pictures taken by the Soviet *Venera* landers confirm that the skies of Venus are a kind of yellow-orange.

Mars is a different story. It is a smaller world than Earth, with a much thinner atmosphere. The pressure at the surface of Mars is,

The first color photo taken from the surface of Mars erroneously showed an earthly blue sky (top). Using correct color calibration of the spacecraft cameras, a much redder sky was revealed (bottom). Courtesy JPL/NASA.

in fact, about the same as the altitude in the Earth's stratosphere to which Simons rose. So we might expect the Martian sky to be black or purple-black. The first color picture from the surface of Mars was obtained in July 1976 by the American *Viking 1* lander—the first spacecraft to touch down successfully on the surface of the Red Planet. The digital data were dutifully radioed from

Mars back to Earth, and the color picture assembled by computer. To the surprise of all the scientists and nobody else, that first image, released to the press, showed the Martian sky to be a comfortable, homey blue—impossible for a planet with so insubstantial an atmosphere. Something had gone wrong.

The picture on your color television set is a mixture of three monochrome images, each in a different color of light—red, green, and blue. You can see this method of color compositing in video projection systems, which project separate beams of red,

The skies of four terrestrial planets—Mercury, Venus, Earth, and Mars—as depicted by artist Don Davis.

Changes in sky color on Mars.
In these pictures, taken by the
camera on *Viking Lander 1*, the
sky colors can be compared day
by day. Around mission day
1,742 a great dust storm erupted,
darkening and reddening the sky.
Viking images, courtesy
IPl /NASA.

green, and blue light to generate a full-color picture (including
yellows). To get the right color, your set needs to mix or balance
these three monochrome images correctly. If you turn up the intensity of, say, blue, the picture will appear too blue. Any picture
returned from space requires a similar color balance. Considerable
discretion is sometimes left to the computer analysts in deciding
this balance. The *Viking* analysts were not planetary astronomers,
and with this first color picture from Mars they simply mixed the
colors until it looked "right." We are so conditioned by our experience on Earth that "right," of course, means a blue sky. The color
of the picture was soon corrected—using color calibration standards placed for this very purpose on board the spacecraft—and
the resulting composite showed no blue sky at all; rather it was
something between ochre and pink. Not blue, but hardly purple-
black either.

This *is* the right color of the Martian sky. Much of the surface
of Mars is desert—and red because the sands are rusty. There are
occasional violent sandstorms that lift fine particles from the surface high into the atmosphere. It takes a long time for them to fall
out, and before the sky has fully cleaned itself, there's always another sandstorm. Global or near-global sandstorms occur almost
every Martian year. Since rusty particles are always suspended in
this sky, future generations of humans, born and living out their
lives on Mars, will consider that salmon color to be as natural and
familiar as we consider our homey blue. From a single glance at
the daytime sky, they'll probably be able to tell how long it's been
since the last big sandstorm.

The planets in the outer Solar System—Jupiter, Saturn,
Uranus, and Neptune—are of a different sort. These are huge
worlds with giant atmospheres made mainly of hydrogen and helium. Their solid surfaces are so deep inside that no sunlight pene-
trates there at all. Down there, the sky is black, with no prospect of
a sunrise—not ever. The perpetual starless night is perhaps illuminated on occasion by a bolt of lightning. But higher in the atmosphere, where the sunlight reaches, a much more beautiful vista
awaits.

On Jupiter, above a high-altitude haze layer composed of

ammonia (rather than water) ice particles, the sky is almost black. Farther down, in the blue sky region, are multicolored clouds—in various shades of yellow-brown, and of unknown composition. (The candidate materials include sulfur, phosphorus, and complex organic molecules.) Even farther down, the sky will appear redbrown, except that the clouds there are of varying thicknesses, and where they are thin, you might see a patch of blue. Still deeper, we gradually return to perpetual night. Something similar is true on Saturn, but the colors there are much more muted.

Uranus and especially Neptune have an uncanny, austere blue color through which clouds—some of them a little whiter—are carried by high-speed winds. Sunlight reaches a comparatively clean atmosphere composed mainly of hydrogen and helium but also rich in methane. Long paths of methane absorb yellow and especially red light and let the green and blue filter through. A thin hydrocarbon haze removes a little blue. There may be a depth where the sky is greenish.

Conventional wisdom holds that the absorption by methane and the Rayleigh scattering of sunlight by the deep atmosphere together account for the blue colors on Uranus and Neptune. But analysis of *Voyager* data by Kevin Baines of JPL seems to show that these causes are insufficient. Apparently very deep—maybe in the vicinity of hypothesized clouds of hydrogen sulfide—there is an abundant blue substance. So far no one has been able to figure out what it might be. Blue materials are very rare in Nature. As always happens in science, the old mysteries are dispelled only to be replaced by new ones. Sooner or later we'll find out the answer to this one, too.

ALL WORLDS WITH NONBLACK SKIES have atmospheres. If you're standing on the surface and there's an atmosphere thick enough to see, there's probably a way to fly through it. We're now sending our instruments to fly in the variously colored skies of other worlds. Someday we will go ourselves.

Parachutes have already been used in the atmospheres of Venus and Mars, and are planned for Jupiter and Titan. In 1985 two French-Soviet balloons sailed through the yellow skies of

The skies of three planets: ABOVE, Venus; NEXT PAGE, Jupiter and Uranus. No direct measurements of cloud composition for any of the Jovian planets can be made prior to the *Galileo* entry probe into Jupiter. Paintings by Don Davis.

Venus. The *Vega 1* balloon, about 4 meters across, dangled an instrument package 13 meters below. The balloon inflated in the night hemisphere, floated about 54 kilometers above the surface, and transmitted data for almost two Earth days before its batteries failed. In that time it traveled 11,600 kilometers (nearly 7,000 miles) over the surface of Venus, far below. The *Vega 2* balloon had an almost identical profile. The atmosphere of Venus has also been used for aerobraking—changing the *Magellan* spacecraft's orbit by friction with the dense air; this is a key future technology for converting flyby spacecraft to Mars into orbiters and landers.

A Mars mission, scheduled to be launched in 1998, and led by Russia, includes an enormous French hot air balloon—looking something like a vast jellyfish, a Portuguese man-of-war. It's designed to sink to the Martian surface every chilly twilight and rise high when heated by sunlight the next day. The winds are so fast that, if all goes well, it will be carried hundreds of kilometers each day, hopping and skipping over the north pole. In the early morning, when close to the ground, it will obtain very high resolution pictures and other data. The balloon has an instrumental guiderope, essential for its stability, conceived and designed by a private membership organization based in Pasadena, California, The Planetary Society.

Since the surface pressure on Mars is approximately that at an altitude of 100,000 feet on Earth, we know we can fly airplanes there. The U–2, for example, or the SR–71 *Blackbird* routinely approaches such low pressures. Aircraft with even larger wingspans have been designed for Mars.

The dream of flight and the dream of space travel are twins, conceived by similar visionaries, dependent on allied technologies, and evolving more or less in tandem. As certain practical and economic limits to flight on Earth are reached, the possibility arises of flying through the multihued skies of other worlds.

IT IS NOW ALMOST POSSIBLE to assign color combinations, based on the colors of clouds and sky, to every planet in the Solar System—from the sulfur-stained skies of Venus and the rusty skies of Mars to the aquamarine of Uranus and the hypnotic and unearthly blue

The French Mars balloon settles down on the Martian surface at evening, trailing its instrumented guide rope designated "SNAKE." This balloon may be launched as part of the Russian 1998 Mars mission. Painting by Michael Carroll.

of Neptune. *Sacre-jaune, sacre-rouge, sacre-vert.* Perhaps they will one day adorn the flags of distant human outposts in the Solar System, in that time when the new frontiers are sweeping out from the Sun to the stars, and the explorers are surrounded by the endless black of space. *Sacre-noir.*

EVENING AND MORNING STAR

> This is another world
> Which is not of men.
> —LI BAI, "QUESTION AND ANSWER IN THE MOUNTAINS"
> (CHINA, TANG DYNASTY, CA. 730)

Y ou can see it shining brilliantly in the twilight, chasing the Sun down below the western horizon. Upon first glimpsing it each night, people were accustomed to make a wish ("upon a star"). Sometimes the wish came true.

Or you can spy it in the east before dawn, fleeing the rising Sun. In these two incarnations, brighter than anything else in the sky except only the Sun and the Moon, it was known as the evening and the morning star. Our ancestors did not recognize it was a world, the same world, never too far from the Sun because it

OPPOSITE: A *Mariner*-class spacecraft in interplanetary space. *Mariner 2* in 1962 was the first successful interplanetary mission and the first successful Venus probe. It opened the age of planetary exploration. Courtesy JPL.

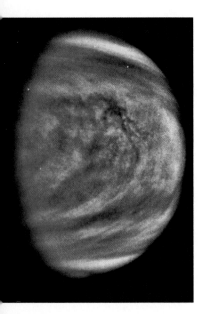

The upper clouds of Venus as imaged by the *Galileo* spacecraft. The rendition has been colorized blue to bring out subtle contrasts (and to indicate that it was taken through a violet filter). These sulfuric acid clouds are highly convective; it would be a very bumpy ride flying an airplane through them. No hint of surface features can be made out. Courtesy JPL/NASA.

is in an orbit about it interior to the Earth's. Just before sunset or just after sunrise, we can sometimes see it near some fluffy white cloud, and then discover by the comparison that Venus has a color, a pale lemon-yellow.

You peer through the eyepiece of a telescope—even a big telescope, even the largest optical telescope on Earth—and you can make out no detail at all. Over the months, you see a feature-less disk methodically going through phases, like the Moon: crescent Venus, full Venus, gibbous Venus, new Venus. There is not a hint of continents or oceans.

Some of the first astronomers to see Venus through the telescope immediately recognized that they were examining a world enshrouded by clouds. The clouds, we now know, are droplets of concentrated sulfuric acid, stained yellow by a little elemental sulfur. They lie high above the ground. In ordinary visible light there's no hint of what this planet's surface, some 50 kilometers below the cloud tops, is like, and for centuries the best we had were wild guesses.

You might conjecture that if we could take a much finer look there might be breaks in the clouds, revealing day by day, in bits and pieces, the mysterious surface ordinarily hidden from our view. Then the time of guesses would be over. The Earth is on average half cloud-covered. In the early days of Venus exploration, we saw no reason that Venus should be 100 percent overcast. If instead it was only 90 percent, or even 99 percent, cloud-covered, the transient patches of clearing might tell us much.

In 1960 and 1961, *Mariners 1* and *2,* the first American spacecraft designed to visit Venus, were being prepared. There were those, like me, who thought the ships should carry video cameras so they could radio pictures back to Earth. The same technology would be used a few years later when *Rangers 7, 8,* and *9* would photograph the Moon on the way to their crash landings—the last making a bull's-eye in the crater Alphonsus. But time was short for the Venus mission, and cameras were heavy. There were those who maintained that cameras weren't really scientific instruments, but rather catch-as-catch-can, razzle-dazzle, pandering to the public, and unable to answer a single straightforward, well-posed scientific question. I thought myself that whether there are breaks in the clouds was one such question. I argued that cameras could also answer questions that we were too dumb even to pose. I argued that pictures were the only way to show the public—who were, after all, footing the bill—the excitement of robotic missions. At any rate, no camera was flown, and subsequent missions have, for this particular world, at least partly vindicated that judgment: Even at high resolution from close flybys, in visible light it turns out there are no breaks in the clouds of Venus, any more than in the clouds of Titan.* These worlds are permanently overcast.

In the ultraviolet there is detail, but due to transient patches of high-altitude overcast, far above the main cloud deck. The high clouds race around the planet much faster than the planet itself turns: super-rotation. We have an even smaller chance of seeing the surface in the ultraviolet.

When it became clear that the atmosphere of Venus was much thicker than the air on Earth—as we now know, the pressure at the surface is ninety times what it is here—it immediately followed that in ordinary visible light we could not possibly see the surface, even if there *were* breaks in the clouds. What little sunlight is able to make its tortuous way through the dense atmosphere to the surface would be reflected back, all right; but the photons would be so jumbled by repeated scattering off molecules in the lower air that no image of surface features could be retained. It would be like a "whiteout" in a polar snowstorm. However, this effect, intense Rayleigh scattering, declines rapidly with increasing wavelength; in the near-infrared, it was easy to calculate, you *could* see the surface if there were breaks in the clouds—or if the clouds were transparent there.

So in 1970 Jim Pollack, Dave Morrison, and I went to the Mc-

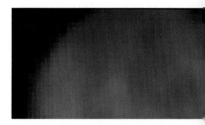

Ground-based near-infrared photography of Venus, just barely revealing surface features under the clouds. Courtesy Anglo-Australian Observatory. Photograph by David Allen.

* For Titan, imaging revealed a succession of detached hazes above the main layer of aerosols. So Venus works out to be the only world in the Solar System for which spacecraft cameras working in ordinary visible light *haven't* discovered something important. Happily, we've now returned pictures from almost every world we've visited. (NASA's *International Cometary Explorer,* which raced through the tail of Comet Giacobini-Zimmer in 1985, flew blind, being devoted to charged particles and magnetic fields.)

Donald Observatory of the University of Texas to try to observe Venus in the near-infrared. We "hypersensitized" our emulsions; the good old-fashioned[*] glass photographic plates were treated with ammonia, and sometimes heated or briefly illuminated, before being exposed at the telescope to light from Venus. For a time the cellars of McDonald Observatory reeked of ammonia. We took many pictures. None showed any detail. We concluded that either we hadn't gone far enough into the infrared, or the clouds of Venus were opaque and unbroken in the near infrared.

More than 20 years later, the *Galileo* spacecraft, making a close flyby of Venus, examined it with higher resolution and sensitivity, and at wavelengths a little further into the infrared than we were able to reach with our crude glass emulsions. *Galileo* photographed great mountain ranges. We already knew of their existence, though; a much more powerful technique had earlier been employed: radar. Radio waves effortlessly penetrate the clouds and thick atmosphere of Venus, bounce off the surface, and return to Earth, where they are gathered in and used to make a picture. The

Galileo observes the topography of Venus through the dense clouds. This color image is constructed from observations at 1.18, 1.74, and 2.3 micrometers in the near-infrared part of the spectrum. White is high; blue is low. The observations are compared with what the *Pioneer 12* radar instrument found. *Galileo*/NIMS data, courtesy Robert Carlson of JPL and NASA.

---

[*] Today many telescopic images are obtained with such electronic contrivances as charge-coupled devices and diode arrays, and processed by computer—all technologies unavailable to astronomers in 1970.

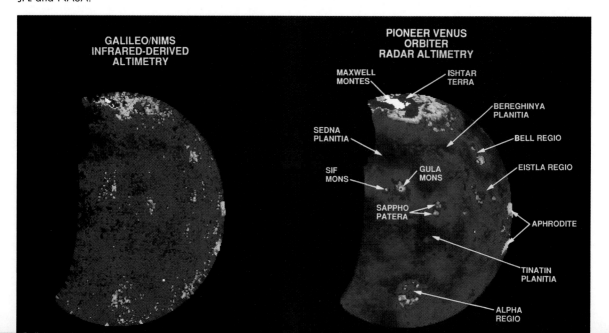

first work had been done, chiefly, by American ground-based radar at JPL's Goldstone tracking station in the Mojave Desert and at the Arecibo Observatory in Puerto Rico, operated by Cornell University.

Then the U.S. *Pioneer 12*, the Soviet *Venera 15* and *16* and the U.S. *Magellan* missions inserted radar telescopes into orbit around Venus and mapped the place pole to pole. Each spacecraft would transmit a radar signal to the surface and then catch it as it bounced back. From how reflective each patch of surface was and how long it took the signal to return (shorter from mountains, longer from valleys), a detailed map of the entire surface was slowly and painstakingly constructed.

The world so revealed turns out to be uniquely sculpted by lava flows (and, to a much lesser degree, by wind), as described in the next chapter. The clouds and atmosphere of Venus have now become transparent to us, and another world has been visited by the doughty robot explorers from Earth. Our experience with Venus is now being applied elsewhere—especially to Titan, where once again impenetrable clouds hide an enigmatic surface, and radar is beginning to give us hints of what might lie below.

VENUS HAD LONG BEEN THOUGHT of as our sister world. It is the nearest planet to the Earth. It has almost the same mass, size, density, and gravitational pull as the Earth does. It's a little closer to the Sun than the Earth, but its bright clouds reflect more sunlight back to space than our clouds do. As a first guess you might very well imagine that, under those unbroken clouds, Venus was rather like Earth. Early scientific speculation included fetid swamps crawling with monster amphibians, like the Earth in the Carboniferous Period; a world desert; a global petroleum sea; and a seltzer ocean dotted here and there with limestone-encrusted islands. While based on some scientific data, these "models" of Venus—the first dating from the beginnings of the century, the second from the 1930s, and the last two from the mid-1950s—were little more than scientific romances, hardly constrained by the sparse data available.

Then, in 1956, a report was published in *The Astrophysical*

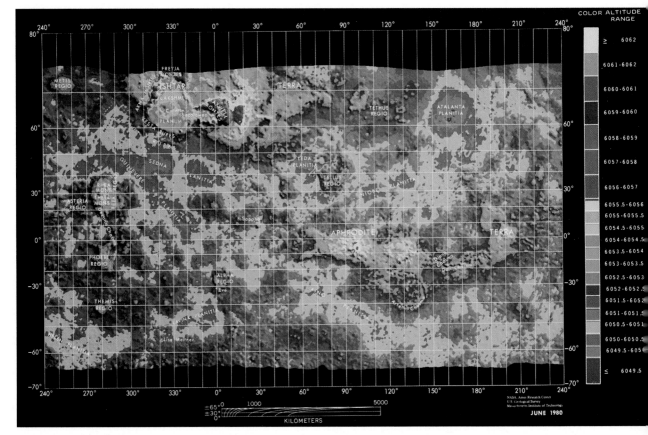

Early radar map of Venus, relying on ground-based and *Pioneer 12* observations through the clouds. Color scale is at right: white, pink, and red are high; blue and purple are low. Note in particular the region Lakshmi Planum at 65° north latitude, 330° longitude, (upper left). Courtesy USGS/NASA.

Journal by Cornell H. Mayer and his colleagues. They had pointed a newly completed radio telescope, built in part for classified research, on the roof of the Naval Research Laboratory in Washington, D.C., at Venus and measured the flux of radio waves arriving at Earth. This was not radar: No radio waves were bounced off Venus. This was listening to radio waves that Venus on its own emits to space. Venus turned out to be much brighter than the background of distant stars and galaxies. This in itself was not very surprising. Every object warmer than absolute zero (−273°C) gives off radiation throughout the electromagnetic spectrum, including the radio region. You, for example, emit radio waves at an effective or "brightness" temperature of about 35°C, and if you were in surroundings colder than you are, a sensitive radio telescope could detect the faint radio waves you are transmitting in all directions. Each of us is a source of cold static.

What *was* surprising about Mayer's discovery was that the

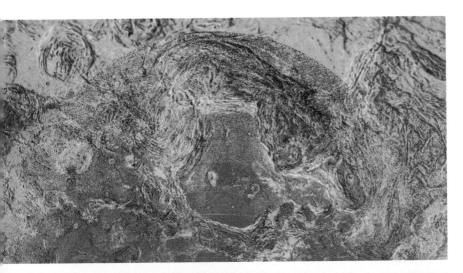

Air-brushed shaded relief maps of Lakshmi Planum on Venus, merging Arecibo Observatory, *Pioneer 12* American spacecraft, and *Venera 15* and *16* Soviet spacecraft data. Lakshmi Planum is about the size of Australia. It sits about five kilometers higher than the average level of this planet's surface, and is fringed by mountains (orange) that rise to above ten kilometers. The gray areas are where the mapmakers had to rely on *Venera 15* and *16* data alone. At bottom, oblique view of Lakshmi Planum from the same data. Images courtesy Alfred McEwen and USGS.

brightness temperature of Venus is more than 300°C, far higher than the surface temperature of the Earth or the measured infrared temperature of the clouds of Venus. Some places on Venus seemed at least 200° hotter than the normal boiling point of water. What could this mean?

Soon there was a deluge of explanations. I argued that the high radio brightness temperature was a direct indication of a hot surface, and that the high temperatures were due to a massive carbon dioxide/water vapor greenhouse effect—in which some sunlight is transmitted through the clouds and heats the surface, but the surface experiences enormous difficulty in radiating back to space because of the high infrared opacity of carbon dioxide and water vapor. Carbon dioxide absorbs at a range of wavelengths through the infrared, but there seemed to be "windows" between the CO_2 absorption bands through which the surface could read-

Magellan image in Lakshmi Planum a few hundred kilometers across. The ridge of the Danu Montes Mountains can be seen toward bottom left. Courtesy JPL/NASA.

ily cool off to space. Water vapor, though, absorbs at infrared frequencies that correspond in part to the windows in the carbon dioxide opacity. The two gases together, it seemed to me, could pretty well absorb almost all the infrared emission, even if there was very little water vapor—something like two picket fences, the slats of one being fortuitously positioned to cover the gaps of the other.

There was another very different category of explanation, in which the high brightness temperature of Venus had nothing to do with the ground. The surface could still be temperate, clement, congenial. It was proposed that some region in the atmosphere of Venus or in its surrounding magnetosphere emitted these radio waves to space. Electrical discharges between water droplets in the Venus clouds were suggested. A glow discharge in which ions and electrons recombined at twilight and dawn in the upper atmosphere was offered. A very dense ionosphere had its advocates, in which the mutual acceleration of unbound electrons ("free-free emission") gave off radio waves. (One proponent of this idea even suggested that the high ionization required was due to an average of 10,000 times greater radioactivity on Venus than on Earth—perhaps from a recent nuclear war there.) And, in the light of the discovery of radiation from Jupiter's magnetosphere, it was natural to suggest that the radio emission came from an immense cloud of charged particles trapped by some hypothetical very intense Venusian magnetic field.

In a series of papers I published in the middle 1960s, many in collaboration with Jim Pollack,[*] these conflicting models of a high hot emitting region and a cold surface were subjected to a critical analysis. By then we had two important new clues: the radio spectrum of Venus, and the *Mariner 2* evidence that the radio emission was more intense at the center of the disk of Venus than toward its edge. By 1967 we were able to exclude the alternative models with some confidence, and conclude that the surface of Venus was at a scorching and un-Earthlike temperature, in excess of 400°C. But the argument was inferential, and there were many intermediate steps. We longed for a more direct measurement.

---

[*] James B. Pollack made important contributions to every area of planetary science. He was my first graduate student and a colleague ever since. He converted NASA's Ames Research Center into a world leader in planetary research and the post-doctoral training of planetary scientists. His gentleness was as extraordinary as his scientific abilities. He died in 1994 at the height of his powers.

In October 1967—commemorating the tenth anniversary of *Sputnik 1*—the Soviet *Venera 4* spacecraft dropped an entry capsule into the clouds of Venus. It returned data from the hot lower atmosphere, but did not survive to the surface. One day later, the United States spacecraft *Mariner 5* flew by Venus, its radio transmission to Earth skimming the atmosphere at progressively greater depths. The rate of fading of the signal gave information about atmospheric temperatures. Although there seemed to be some discrepancies (later resolved) between the two sets of spacecraft data, both clearly indicated that the surface of Venus is very hot.

Since then a progression of Soviet *Venera* spacecraft and one cluster of American spacecraft from the *Pioneer 12* mission have entered the deep atmosphere or landed on the surface and measured directly—essentially by sticking out a thermometer—the surface and near-surface temperatures. They turn out to be about 470°C, almost 900°F. When such factors as calibration errors of terrestrial radio telescopes and surface emissivity are taken into account, the old radio observations and the new direct spacecraft measurements turn out to be in good accord.

Early Soviet landers were designed for an atmosphere somewhat like our own. They were crushed by the high pressures like a tin can in the grasp of a champion arm wrestler, or a World War II submarine in the Tonga Trench. Thereafter, Soviet Venus entry vehicles were heavily reinforced, like modern submarines, and successfully landed on the searing surface. When it became clear how

A panorama of frozen lava flows—the basaltic surface of Venus as seen by the *Venera 14* lander. This is roughly what you would see if you were lying prone on the planet's surface. At top left and right are small yellow triangles—a glimpse of the sky of Venus. Courtesy Vernadsky Institute, Moscow.

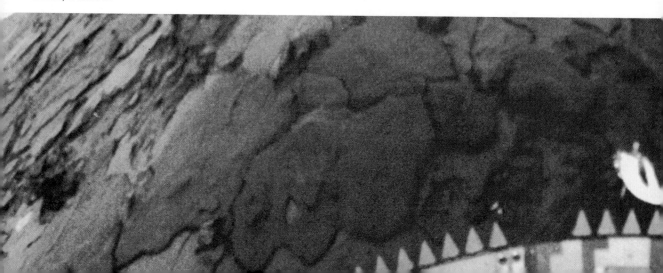

deep the atmosphere is and how thick the clouds, Soviet designers became concerned that the surface might be pitch-black. *Veneras 9* and *10* were equipped with floodlights. They proved unnecessary. A few percent of the sunlight that falls on the top of the clouds makes it through to the surface, and Venus is about as bright as on a cloudy day on Earth.

The resistance to the idea of a hot surface on Venus can, I suppose, be attributed to our reluctance to abandon the notion that the nearest planet is hospitable for life, for future exploration, and perhaps even, in the longer term, for human settlement. As it turns out there are no Carboniferous swamps, no global oil or seltzer oceans. Instead, Venus is a stifling, brooding inferno. There are some deserts, but it's mainly a world of frozen lava seas. Our hopes are unfulfilled. The call of this world is now more muted than in the early days of spacecraft exploration, when almost anything was possible and our most romantic notions about Venus might, for all we then knew, be realized.

MANY SPACECRAFT CONTRIBUTED to our present understanding of Venus. But the pioneering mission was *Mariner 2. Mariner 1* failed at launch and—as they say of a racehorse with a broken leg—had to be destroyed. *Mariner 2* worked beautifully and provided the key early radio data on the climate of Venus. It made infrared observations of the properties of the clouds. On its way from Earth to Venus, it discovered and measured the solar wind—the stream of

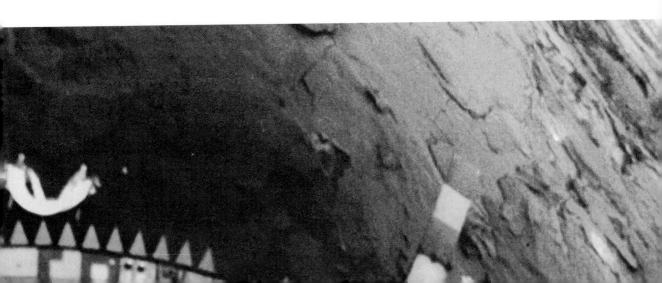

Magellan spacecraft data on the surface of Venus. The colors in the middle image denote altitude; in the first and last, they indicate the efficiency with which the surface reflects radar. The dark straight lines represent regions where no data were acquired. Courtesy USGS.

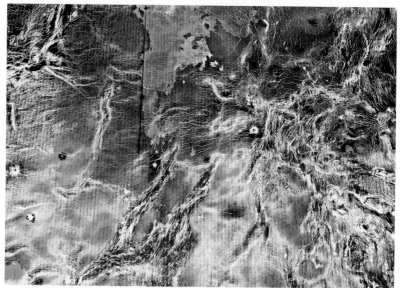

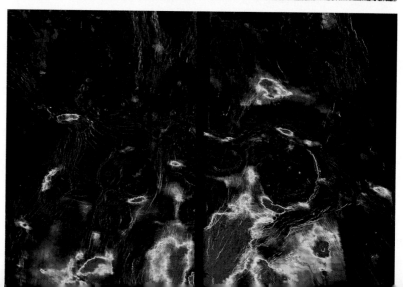

charged particles that flows outward from the Sun, filling the magnetospheres of any planets in its way, blowing back the tails of comets, and establishing the distant heliopause. *Mariner 2* was the first successful planetary probe, the ship that ushered in the age of planetary exploration.

It's still in orbit around the Sun, every few hundred days still approaching, more or less tangentially, the orbit of Venus. Each time that happens, Venus isn't there. But if we wait long enough, Venus will one day be nearby and *Mariner 2* will be accelerated by the planet's gravity into some quite different orbit. Ultimately, *Mariner 2,* like some planetesimal from ages past, will be swept up by another planet, fall into the Sun, or be ejected from the Solar System.

Until then, this harbinger of the age of planetary exploration, this minuscule artificial planet, will continue silently orbiting the Sun. It's a little as if Columbus's flagship, the *Santa María,* were still making regular runs with a ghostly crew across the Atlantic between Cádiz and Hispaniola. In the vacuum of interplanetary space, *Mariner 2* should be in mint condition for many generations.

My wish on the evening and morning star is this: that late in the twenty-first century some great ship, on its regular gravity-assisted transit to the outer Solar System, intercepts this ancient derelict and heaves it aboard, so it can be displayed in a museum of early space technology—on Mars, perhaps, or Europa, or Iapetus.

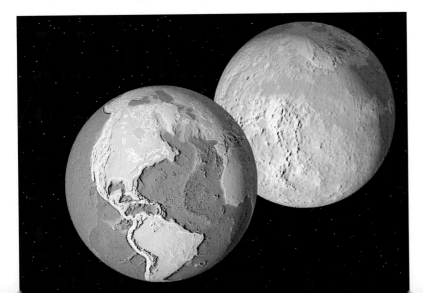

Sister worlds: the Earth denuded of oceans and Venus of its dense atmosphere. They may have begun under very similar circumstances, but the two planets have evolved in very different directions. Image courtesy JPL/NASA.

THE GROUND MELTS

Midway between Thera and Therasia, fires broke forth from the sea and
continued for four days, so that the whole sea boiled and blazed, and the
fires cast up an island which was gradually elevated as though by levers . . .
After the cessation of the eruption, the Rhodians, at the time
of their maritime supremacy, were first to venture upon
the scene and to erect on the island a temple.

STRABO, *GEOGRAPHY* (CA. 7 B.C.)

All over the Earth, you can find a kind of mountain with one striking and unusual feature. Any child can recognize it: The top seems sheared or squared off. If you climb to the summit or fly over it, you discover that the mountain has a hole or crater at its peak. In some mountains of this sort, the craters are small; in others, they are almost as big as the mountain itself. Occasionally, the craters are filled with water. Sometimes they're filled with a more amazing liquid: You tiptoe to the edge, and see vast, glowing lakes of yellow-red liquid and fountains of

OPPOSITE: The volcanos of Io greet the *Galileo* spacecraft, December 1995. Jupiter and its thin ring loom in the background. Painting © David A. Hardy.

fire. These holes in the tops of mountains are called calderas, after the word "caldron," and the mountains on which they sit are known, of course, as volcanos—after Vulcan, the Roman god of fire. There are perhaps 600 active volcanos discovered on Earth. Some, beneath the oceans, are yet to be found.

A typical volcanic mountain looks safe enough. Natural vegetation runs up its sides. Terraced fields decorate its flanks. Hamlets and shrines nestle at its base. And yet, without warning, after centuries of lassitude, the mountain may explode. Barrages of boulders, torrents of ash drop out of the sky. Rivers of molten rock come pouring down its sides. All over the Earth people imagined that an active volcano was an imprisoned giant or demon struggling to get out.

The eruptions of Mt. St. Helens and Mt. Pinatubo are recent reminders, but examples can be found throughout history. In 1902 a hot, glowing volcanic cloud swept down the slopes of Mt. Pelée and killed 35,000 people in the city of St. Pierre on the Caribbean island of Martinique. Massive mudflows from the eruption of the Nevado del Ruiz volcano in 1985 killed more than 25,000 Colombians. The eruption of Mt. Vesuvius in the first century buried in ash the hapless inhabitants of Pompeii and Herculaneum and killed the intrepid naturalist Pliny the Elder as he made his way up the side of the volcano, intent on arriving at a better understanding of its workings. (Pliny was hardly the last: Fifteen volcanologists have been killed in sundry volcanic eruptions between 1979 and 1993.) The Mediterranean island of Santorin (also called Thera) is in reality the only part above water of the rim of a volcano now inundated by the sea.★ The explosion of the Santorin volcano in 1623 B.C. may, some historians think, have helped destroy the great Minoan civilization on the nearby island of Crete and changed the balance of power in early classical civilization. This disaster may be the origin of the Atlantis legend as related by Plato, in which a civilization was destroyed "in a single day and night of misfortune." It must have been easy back then to think that a god was angry.

★ The eruption of a nearby submarine volcano and the rapid construction of a new island in 197 B.C. are described by Strabo in the epigraph to this chapter.

Mount St. Helens in the State of Washington, U.S.A., in Landsat 6 false color. Eleven years after the 1980 eruption, the area surrounding the volcano remains desolate (as indicated by the purples). Vegetation is shown in green. Reproduced by permission of Earth Observation Satellite Company, Lanham, Maryland, U.S.A.

Volcanos have naturally been regarded with fear and awe. When medieval Christians viewed the eruption of Mt. Hekla in Iceland and saw churning fragments of soft lava suspended over the summit, they imagined they were seeing the souls of the damned awaiting entrance to Hell. "Fearful howlings, weeping and gnashing of teeth," "melancholy cries and loud wailings" were dutifully reported. The glowing red lakes and sulfurous gases within the Hekla caldera were thought to be a real glimpse into the underworld and a confirmation of folk beliefs in Hell (and, by symmetry, in its partner, Heaven).

A volcano is, in fact, an aperture to an underground realm much vaster than the thin surface layer that humans inhabit, and far more hostile. The lava that erupts from a volcano is liquid

Cross-section of a volcano tapping underground reservoirs of liquid rock. Painting by Kazuaki Iwasaki.

rock—rock raised to its melting point, generally around 1000°C. The lava emerges from a hole in the Earth; as it cools and solidifies, it generates and later remakes the flanks of a volcanic mountain.

The most volcanically active locales on Earth tend to be along ridges on the ocean floor and island arcs—at the junction of two great plates of oceanic crust—either separating from each other, or one slipping under the other. On the seafloor there are long zones of volcanic eruptions—accompanied by swarms of earthquakes and plumes of abyssal smoke and hot water—that we are just beginning to observe with robot and manned submersible vehicles.

Eruptions of lava must mean that the Earth's interior is extremely hot. Indeed, seismic evidence shows that, only a few hundred kilometers beneath the surface, nearly the entire body of the Earth is at least slightly molten. The interior of the Earth is hot, in part, because radioactive elements there, such as uranium, give off heat as they decay; and in part because the Earth retains some of the original heat released in its formation, when many small worlds fell together by their mutual gravity to make the Earth, and when iron drifted down to form our planet's core.

The molten rock, or magma, rises through fissures in the surrounding heavier solid rocks. We can imagine vast subterranean caverns filled with glowing, red, bubbling, viscous liquids that shoot up toward the surface if a suitable channel is by chance provided. The magma, called lava as it pours out of the summit caldera, does indeed arise from the underworld. The souls of the damned have so far eluded detection.

Once the volcano is fully built from successive outpourings, and the lava is no longer spewing up into the caldera, then it becomes just like any other mountain—slowly eroding because of rainfall and windblown debris and, eventually, the movement of continental plates across the Earth's surface. "How many years can a mountain exist before it is washed to the sea?" asked Bob Dylan in the ballad "Blowing in the Wind." The answer depends on which planet we're talking about. For the Earth, it's typically about ten million years. So mountains, volcanic and otherwise, must be built on the same timescale; otherwise the Earth would be everywhere smooth as Kansas.★

Volcanic explosions can punch vast quantities of matter—mainly fine droplets of sulfuric acid—into the stratosphere. There, for a year or two, they reflect sunlight back to space and cool the Earth. This happened recently with the Philippine volcano, Mt. Pinatubo, and disastrously in 1815–16 after the eruption of the Indonesian volcano Mt. Tambora, which resulted in the famine-ridden "year without a summer." A volcanic eruption in Taupo, New Zealand, in the year 177 cooled the climate of the Mediterranean, half a world away, and dropped fine particles onto the Greenland ice cap. The explosion of Mt. Mazama in Oregon (which left the caldera now called Crater Lake) in 4803 B.C. had climatic consequences throughout the northern hemisphere. Studies of volcanic effects on the climate were on the investigative path that eventually led to the discovery of nuclear winter. They provide important tests of our use of computer models to predict future climate change. Volcanic particles injected into the upper air are also an additional cause of thinning of the ozone layer.

★ Even with its mountains and submarine trenches, our planet is astonishingly smooth. If the Earth were the size of a billiard ball, the largest protuberances would be less than a tenth of a millimeter in size—on the threshold of being too small to see or feel.

So a large volcanic explosion in some unfrequented and obscure part of the world can alter the environment on a global scale. Both in their origins and in their effects, volcanos remind us of how vulnerable we are to minor burps and sneezes in the Earth's internal metabolism, and how important it is for us to understand how this subterranean heat engine works.

IN THE FINAL STAGES of formation of the Earth—as well as the Moon, Mars, and Venus—impacts by small worlds are thought to have generated global magma oceans. Molten rock flooded the pre-existing topography. Great floods, tidal waves kilometers high, of flowing, red-hot liquid magma welled up from the interior and poured over the surface of the planet, burying everything in their path: mountains, channels, craters, perhaps even the last evidence of much earlier, more clement times. The geological odometer was reset. All accessible records of surface geology begin with the last global magma flood. Before they cool and solidify, oceans of lava may be hundreds or even thousands of kilometers thick. In our time, billions of years later, the surface of such a world may be quiet, inactive, with no hint of current vulcanism. Or there may be—as on Earth—a few small-scale reminders of an epoch when the entire surface was flooded with liquid rock.

In the early days of planetary geology, ground-based telescopic observations were all the data we had. A fervent debate had been running for half a century on whether the craters of the Moon were due to impacts or volcanos. A few low mounds with summit calderas were found—almost certainly lunar volcanos. But the big craters—bowl- or pan-shaped and sitting on the flat ground and not the tops of mountains—were a different story. Some geologists saw in them similarities with certain highly eroded volcanos on Earth. Others did not. The best counterargument was that we know there are asteroids and comets that fly past the Moon; they must hit it sometimes; and the collisions must make craters. Over the history of the Moon a large number of such craters should have been punched out. So if the craters we see are not due to impacts, where then are the impact craters? We now know from direct laboratory examination of lunar craters that they are almost entirely of impact origin. But 4 billion years

Oceans of molten rock flood the surface of a terrestrial planet early in its history. Painting by Michael Carroll.

ago this little world, nearly dead today, was bubbling and churning away, driven by primeval vulcanism from sources of internal heat now long gone.

In November 1971, NASA's *Mariner 9* spacecraft arrived at Mars to find the planet completely obscured by a global dust storm. Almost the only features to be seen were four circular spots

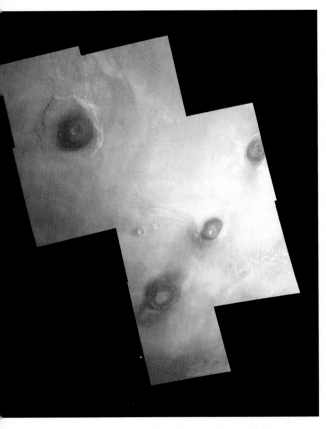

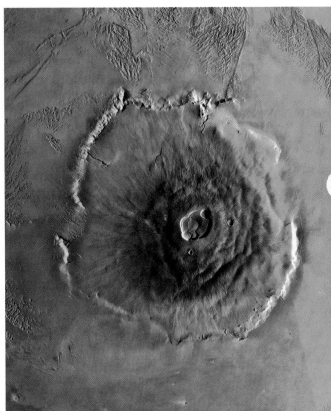

TOP LEFT: The volcanos of the Tharsis Plateau of Mars. RIGHT: Olympus Mons, the largest volcanic construct in the Solar System. Photomosaic from *Viking* data. The summits of these four volcanos with their calderas were all of the surface of Mars that could be seen by the *Mariner 9* spacecraft during the height of the 1971 dust storm. Courtesy USGS/NASA.

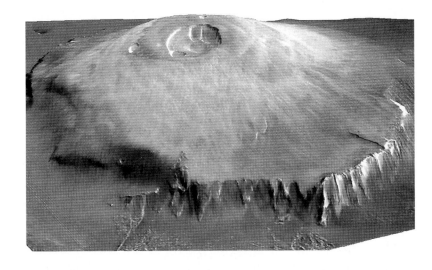

Oblique view of Olympus Mons, reconstructed from *Viking* data. The sparsity of impact craters on the slopes of this great mountain points to its comparative youth. *Viking* photomosaic, courtesy USGS/NASA.

rising out of the reddish murk. But there was something peculiar about them: They had holes in their tops. As the storm cleared, we were able to see unmistakably that we had been viewing four huge volcanic mountains penetrating through the dust cloud, each capped by a great summit caldera.

After the storm dissipated, the true scale of these volcanos became clear. The largest—appropriately named Olympus Mons, or Mt. Olympus, after the home of the Greek gods—is more than 25 kilometers (roughly 15 miles) high, dwarfing not only the largest volcano on Earth but also the largest mountain of any sort, Mt. Everest, which stands 9 kilometers above the Tibetan plateau. There are some 20 large volcanos on Mars, but none so massive as Olympus Mons, which has a volume about 100 times that of the largest volcano on Earth, Mauna Loa in Hawaii.

By counting the accumulated impact craters (made by small impacting asteroids, and readily distinguished from summit

The Arsia Mons volcanic eminence on Mars, with its enormous summit caldera. The map ranges from the Equator to 15° south. USGS shaded relief map.

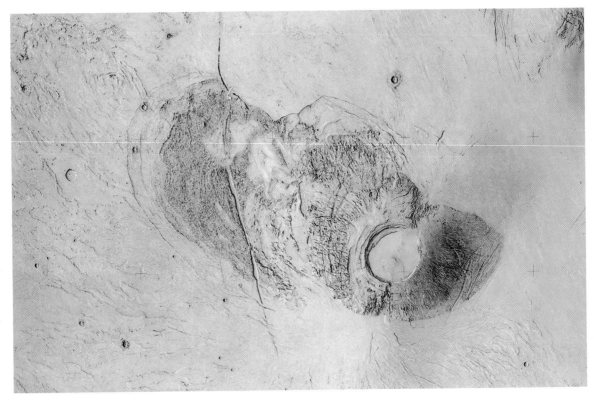

calderas) on the flanks of the volcanos, estimates of their ages can be derived. Some Martian volcanos turn out to be a few billion years old, although none dates back to the very origin of Mars, about 4.5 billion years ago. Some, including Olympus Mons, are comparatively new—perhaps only a few hundred million years old. It is clear that enormous volcanic explosions occurred early in Martian history, perhaps providing an atmosphere much denser than the one Mars holds today. What would the place have looked like if we had visited it then?

Some volcanic flows on Mars (for example, in Cerberus) formed as recently as 200 million years ago. It is, I suppose, even possible—although there is no evidence either way—that Olympus Mons, the largest volcano we know about for certain in the Solar System, will be active again. Volcanologists, a patient sort, would doubtless welcome the event.

In 1990–93 the *Magellan* spacecraft returned surprising radar data about the landforms of Venus. Cartographers prepared maps of almost the entire planet, with fine detail down to about 100 meters, the goal-line-to-goal-line distance in an American football stadium. More data were radioed home by *Magellan* than by all other planetary missions combined. Since much of the ocean floor remains unexplored (except perhaps for still-classified data acquired by the U.S. and Soviet navies), we may know more about the surface topography of Venus than about any other planet, Earth included. Much of the geology of Venus is unlike anything seen on Earth or anywhere else. Planetary geologists have given these landforms names, but that doesn't mean we fully understand how they're formed.

Because the surface temperature of Venus is almost 470°C (900°F), the rocks there are much closer to their melting points than are those at the surface of the Earth. Rocks begin to soften and flow at much shallower depths on Venus than on Earth. This is very likely the reason that many geological features on Venus seem to be plastic and deformed.

The planet is covered by volcanic plains and highland plateaus. The geological constructs include volcanic cones, probable shield volcanos, and calderas. There are many places where we can see that lava has erupted in vast floods. Some plains features ranging to over 200 kilometers in size are playfully called "ticks" and "arachnoids" (which translates roughly as "spiderlike things")—because they are circular depressions surrounded by concentric rings, while long, spindly surface cracks extend radially out from the center. Odd, flat "pancake domes"—a geological feature unknown on Earth, but probably a kind of volcano—are perhaps formed by thick, viscous lava slowly flowing uniformly in all directions. There are many examples of more irregular lava flows. Curious ring structures called "coronae" range up to some 2,000 kilometers across. The distinctive lava flows on stifling hot Venus offer up a rich menu of geological mysteries.

The most unexpected and peculiar features are the sinuous channels—with meanders and oxbows, looking just like river valleys on Earth. The longest are longer than the greatest rivers on

Ridges that have subsequently been flooded by lava in the Ovda Regio highlands of Venus. Vertical relief has been exaggerated 22.5 times for clarity. Reconstructed from *Magellan* data, courtesy JPL/NASA.

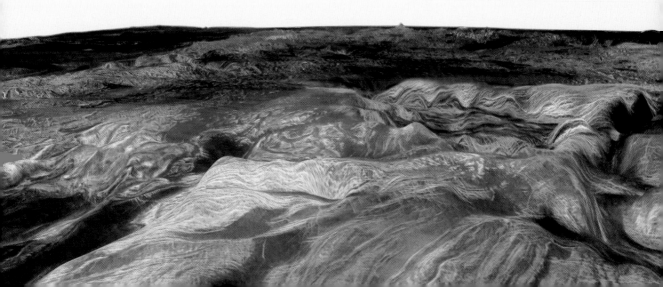

Three (comparatively rare) impact craters on the surface of Venus. Howe is in the foreground and is about 37 kilometers (23 miles) across. Danilova (background left) and Aglaonice (background right) can also be seen. *Magellan* observations, courtesy JPL/NASA.

Magellan views the mountains of Venus. Courtesy JPL/NASA.

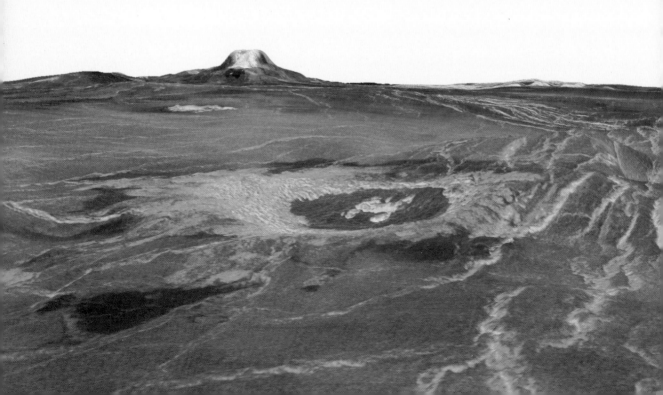

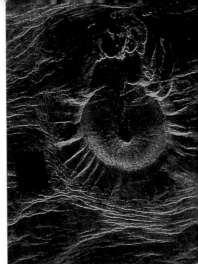

The Maat Mons volcano of Venus. Lava flows extend for hundreds of kilometers across the fractured plains in foreground to the base of this 8-kilometer (5-mile)-high volcano. *Magellan* observations, courtesy JPL/NASA.

An arachnoid. False-color image of a volcanic edifice of a sort unknown on Earth, in Eistla Regio on Venus. It is approximately 66 kilometers (41 miles) across at the base. *Magellan* data, courtesy JPL/NASA.

Earth. But it is much too hot for liquid water on Venus. And we can tell from the absence of small impact craters that the atmosphere has been this thick, driving as great a greenhouse effect, for as long as the present surface has been in existence. (If it had been much thinner, intermediate-sized asteroids would not have burned up on entry into the atmosphere, but would have survived to excavate craters as they impact this planet's surface.) Lava flowing downhill does make sinuous channels (sometimes under the ground, followed by collapse of the roof of the channel). But even at the temperatures of Venus, the lavas radiate heat, cool, slow, congeal, and stop. The magma freezes solid. Lava channels cannot go even 10 percent of the length of the long Venus channels before they solidify. Some planetary geologists think there must be a special thin, watery, inviscid lava generated on Venus. But this is a speculation supported by no other data, and a confession of our ignorance.

The thick atmosphere moves sluggishly; because it's so dense, though, it's very good at lifting and moving fine particles. There are wind streaks on Venus, largely emanating from impact craters, in which the prevailing winds have scoured piles of sand and dust and provided a sort of weather vane imprinted on the surface. Here and there we seem to see fields of sand dunes, and provinces where wind erosion has sculpted volcanic landforms. These aeolian processes take place in slow motion, as if at the bottom of the

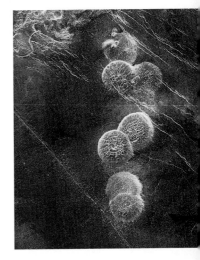

Another domed feature unknown on Earth, also thought to be volcanic. These are called "pancakes." *Magellan* data, courtesy JPL/NASA

Sinuous channel north of the
Fruyja Mountains of Venus.
Magellan data, courtesy
JPL/NASA.

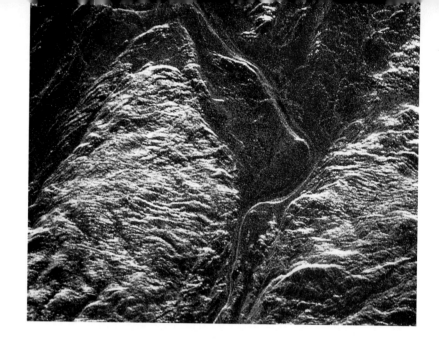

sea. The winds are feeble at the surface of Venus. It may take only a soft gust to raise a cloud of fine particles, but in that stifling inferno a gust is hard to come by.

There are many impact craters on Venus, but nothing like the number on the Moon or Mars. Craters smaller than a few kilometers across are oddly missing. The reason is understood: Small asteroids and comets are broken up on entry into the dense Venus atmosphere before they can hit the surface. The observed cutoff in crater size corresponds very well to the present density of the atmosphere of Venus. Certain irregular splotches seen on the *Magellan* images are thought to be the remains of impactors that broke up in the thick air before they could gouge out a crater.

Most of the impact craters are remarkably pristine and well preserved; only a few percent of them have been engulfed by subsequent lava flows. The surface of Venus as revealed by *Magellan* is very young. There are so few impact craters that everything older than about 500 million years [*] must have been eradicated—on a planet almost certainly 4.5 billion years old. There is only one

[*] The age of the Venus surface, as determined by *Magellan* radar imagery, puts an additional nail in the coffin of the thesis of Immanuel Velikovsky—who around 1950 proposed, to surprising media acclaim, that 3,500 years ago Jupiter spat out a giant "comet" which made several grazing collisions with the Earth, causing various events chronicled in the ancient books of many peoples (such as the Sun standing still on Joshua's command), and then transformed itself into the planet Venus. There are still people who take these notions seriously.

plausible erosive agent adequate for what we see: vulcanism. All over the planet craters, mountains, and other geological features have been inundated by seas of lava that once welled up from the inside, flowed far, and froze.

After examining so young a surface covered with congealed magma, you might wonder if there are any active volcanos left. None has been found for certain, but there are a few—for example, one called Maat Mons—that appear to be surrounded by fresh lava and which may indeed still be churning and belching. There is some evidence that the abundance of sulfur compounds in the high atmosphere varies with time, as if volcanos at the surface were episodically injecting these materials into the atmosphere. When the volcanos are quiescent, the sulfur compounds simply fall out of the air. There's also disputed evidence of lightning playing around the mountaintops of Venus, as sometimes happens on active volcanos on Earth. But we do not know for certain whether there is ongoing vulcanism on Venus. That's a matter for future missions.

Some scientists believe that until about 500 million years ago the Venus surface was almost entirely devoid of landforms. Streams and oceans of molten rock were relentlessly pouring out of the interior, filling in and covering over any relief that had managed to form. Had you plummeted down through the clouds in that long-ago time, the surface would have been nearly uniform and featureless. At night the landscape would have been hellishly glowing from the red heat of molten lava. In this view, the great internal heat engine of Venus, which supplied copious amounts of magma to the surface until about 500 million years ago, has now turned off. The planetary heat engine has finally run down.

In another provocative theoretical model, this one by the geophysicist Donald Turcotte, Venus has plate tectonics like the Earth's—but it turns off and on. Right now, he proposes, the plate tectonics are off; "continents" do not move along the surface, do not crash into one another, do not thereby raise mountain ranges, and are not later subducted into the deep interior. After hundreds of millions of years of quiescence, though, plate tectonics always breaks out and surface features are flooded by lava, destroyed by mountain building, subducted, and otherwise obliterated. The last such breakout ended about 500 million years ago, Turcotte suggests, and everything has been quiet since. However, the presence of coronae may signify—on timescales that are geologically in the near future—that massive changes on the surface of Venus are about to break out again.

EVEN MORE UNEXPECTED than the great Martian volcanos or the magma-flooded surface of Venus is what awaited us when the *Voyager 1* spacecraft encountered Io, the innermost of the four large Galilean moons of Jupiter, in March 1979. There we found a strange, small, multihued world positively awash in volcanos. As we watched in astonishment, eight active plumes poured gas and fine particles up into the sky. The largest, now called Pelé—after the Hawaiian volcano goddess—projected a fountain of material 250 kilometers into space, higher above the surface of Io than some astronauts have ventured above the Earth. By the time *Voyager 2* arrived at Io, four months later, Pelé had turned itself off, although six of the other plumes were still active, at least one new plume had been discovered, and another caldera, named Surt, had changed its color dramatically.

The colors of Io, even though exaggerated in NASA's color-enhanced images, are like none elsewhere in the Solar System. The currently favored explanation is that the Ionian volcanos are dri-

BELOW LEFT: Looking down the mouth of Loki Patera (lower center of image), Io. *Voyager* data, image courtesy USGS/NASA.
BELOW RIGHT: The plume from the active volcano Loki Patera on the horizon of Io. False-color *Voyager* data, courtesy USGS/NASA.

ven not by upwelling molten rock, as on the Earth, the Moon, Venus, and Mars, but by upwelling sulfur dioxide and molten sulfur. The surface is covered with volcanic mountains, volcanic calderas, vents, and lakes of molten sulfur. Various forms and compounds of sulfur have been detected on the surface of Io and in nearby space—the volcanos blow some of the sulfur off Io altogether.* These findings have suggested to some an underground sea of liquid sulfur that issues to the surface at points of weakness, generates a shallow volcanic mound, trickles downhill, and freezes, its final color determined by its temperature on eruption.

On the Moon or Mars, you can find many places that have changed little in a billion years. On Io, in a century, much of the surface should be reflooded, filled in or washed away by new volcanic flows. Maps of Io will then quickly become obsolete, and cartography of Io will have become a growth industry.

* Io's volcanos are also the copious source of electrically charged atoms such as oxygen and sulfur that populate a ghostly, doughnut-shaped tube of matter that surrounds Jupiter.

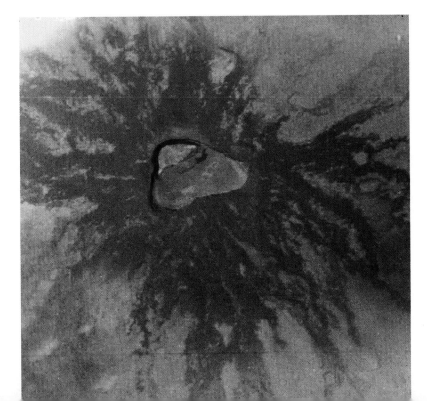

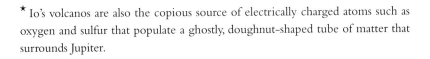

Streamers from the Prometheus volcano on Io shooting into the thin atmosphere and falling back to the ground. *Voyager* data, printed here as negative to bring out contrast. Courtesy USGS/NASA.

The Maasaw Patera volcanic construct on Io. Lava flows, perhaps of molten sulfur, once poured out of the summit caldera. *Voyager* data, courtesy USGS/NASA.

False-color image of the region of
the south pole of Io. *Voyager*
data, image courtesy
USGS/NASA.

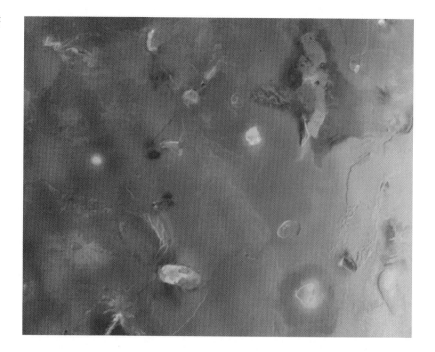

All this seems to follow readily enough from the *Voyager* observations. The rate at which the surface is covered over by current
volcanic flows implies massive changes in 50 or 100 years, a prediction that luckily can be tested. The *Voyager* images of Io can be
compared with much poorer images taken by ground-based telescopes 50 years earlier, and by the Hubble Space Telescope 13
years later. The surprising conclusion seems to be that the big surface markings on Io have hardly changed at all. Clearly, we're missing something.

A VOLCANO in one sense represents the insides of a planet gushing
out, a wound that eventually heals itself by cooling, only to be replaced by new stigmata. Different worlds have different insides.
The discovery of liquid-sulfur vulcanism on Io was a little like
finding that an old acquaintance, when cut, bleeds green. You had
no idea such differences were possible. He seemed so ordinary.

   We are naturally eager to find additional signs of vulcanism
on other worlds. On Europa, the second of the Galilean moons of
Jupiter and Io's neighbor, there are no volcanic mountains at all;
but molten ice—liquid water—seems to have gushed to the sur-

face through an enormous number of crisscrossing dark markings before freezing. And farther out, among the moons of Saturn, there are signs that liquid water has gushed up from the interior and wiped away impact craters. Still, we have never seen anything that might plausibly be an ice volcano in either the Jupiter or Saturn systems. On Triton, we may have observed nitrogen or methane vulcanism.

The volcanos of other worlds provide a stirring spectacle. They enhance our sense of wonder, our joy in the beauty and diversity of the Cosmos. But these exotic volcanos perform another service as well: They help us to know the volcanos of our own world—and perhaps will help one day even to predict their eruptions. If we cannot understand what's happening in other circumstances, where the physical parameters are different, how deep can our understanding be of the circumstance of most concern to us? A general theory of vulcanism must cover all cases. When we stumble upon vast volcanic eminences on a geologically quiet Mars; when we discover the surface of Venus wiped clean only yesterday by floods of magma; when we find a world melted not by the heat of radioactive decay, as on Earth, but by gravitational tides exerted by nearby worlds; when we observe sulfur rather than silicate vulcanism; and when we begin to wonder, in the moons of the outer planets, whether we might be viewing water, ammonia, nitrogen, or methane vulcanism—then we are learning what else is possible.

A large flooded basin on Neptune's moon Triton, about 200 kilometers (120 miles) wide and 400 kilometers (240 miles) long. The material that caused the flooding is not known directly, but is believed to be nitrogen or methane (or just possibly water) ice heated in the interior, rushing up through cracks to the surface, flowing, and then freezing—all in close analogy to what happens with molten lava on Earth. *Voyager 2* image, courtesy JPL/NASA.

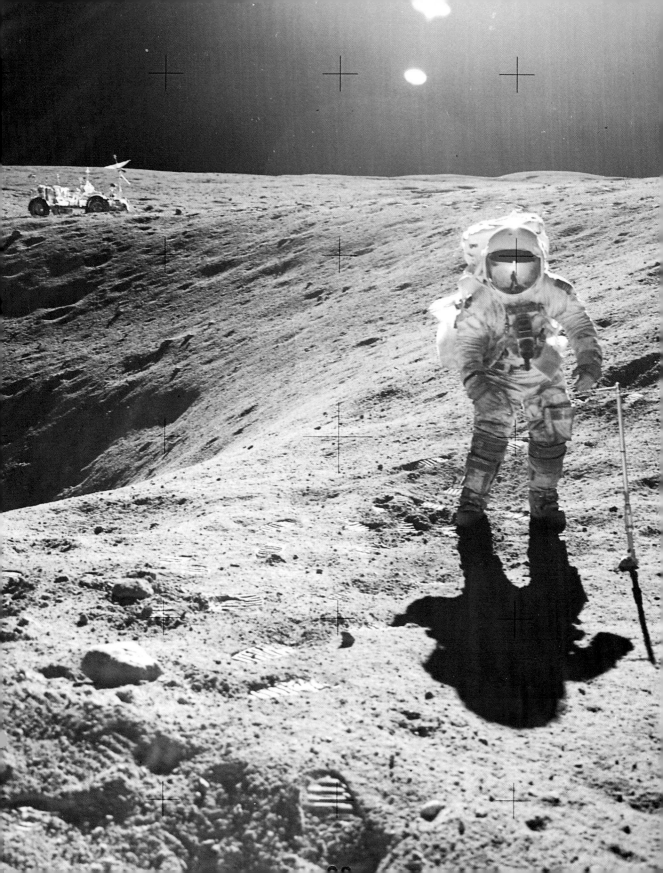

THE GIFT OF *APOLLO*

The gates of Heaven are open wide;
Off I ride . . .

—CH'U TZ'U (ATTRIBUTED TO CH'Ü YÜAN),
"THE NINE SONGS," SONG V, "THE GREAT LORD OF LIVES"
(CHINA, CA. THIRD CENTURY B.C.)

I t's a sultry night in July. You've fallen asleep in the armchair. Abruptly, you startle awake, disoriented. The television set is on, but not the sound. You strain to understand what you're seeing. Two ghostly white figures in coveralls and helmets are softly dancing under a pitch-black sky. They make strange little skipping motions, which propel them upward amid barely percep-tible clouds of dust. But something is wrong. They take too long to come down. Encumbered as they are, they seem to be flying—a little. You rub your eyes, but the dreamlike tableau persists.

Of all the events surrounding *Apollo 11*'s landing on the Moon on July 20, 1969, my most vivid recollection is its unreal

OPPOSITE: An *Apollo* astronaut poses for his portrait on the surface of the Moon. The photographer is mirrored in his visor. The Rover is parked on the far rim of the impact crater at left. Among the very small impact craters at foreground is an astronaut's bootprint. Pictured here is the *Apollo 16* astronaut Charles Duke. Courtesy NASA.

quality. Neil Armstrong and Buzz Aldrin shuffled along the gray, dusty lunar surface, the Earth looming large in their sky, while Michael Collins, now the Moon's own moon, orbited above them in lonely vigil. Yes, it was an astonishing technological achievement and a triumph for the United States. Yes, the astronauts displayed death-defying courage. Yes, as Armstrong said as he first

Apollo 11 liftoff. Courtesy NASA.

Third stage separation of the *Saturn* booster in the 1968 *Apollo 7* mission. Courtesy NASA.

alighted, this was a historic step for the human species. But if you turned off the byplay between Mission Control and the Sea of Tranquility, with its deliberately mundane and routine chatter, and stared into that black-and-white television monitor, you could glimpse that we humans had entered the realm of myth and legend.

We knew the Moon from our earliest days. It was there when our ancestors descended from the trees into the savannahs, when we learned to walk upright, when we first devised stone tools, when we domesticated fire, when we invented agriculture and built cities and set out to subdue the Earth. Folklore and popular songs celebrate a mysterious connection between the Moon and love. The word "month" and the second day of the week are both

A footprint on the Moon. If undisturbed by visitors, this token of the *Apollo* expeditions will last a million years or more. Courtesy NASA.

named after the Moon. Its waxing and waning—from crescent to full to crescent to new—was widely understood as a celestial metaphor of death and rebirth. It was connected with the ovulation cycle of women, which has nearly the same period—as the word "menstruation" (Latin *mensis* = month, from the word "to measure") reminds us. Those who sleep in moonlight go mad; the connection is preserved in the English word "lunatic." In the old Persian story, a vizier renowned for his wisdom is asked which is more useful, the Sun or the Moon. "The Moon," he answers, "because the Sun shines in daytime, when it's light out anyway." Especially when we lived out-of-doors, it was a major—if oddly intangible—presence in our lives.

The Moon was a metaphor for the unattainable: "You might as well ask for the Moon," they used to say. Or "You can no more do that than fly to the Moon." For most of our history, we had no idea what it was. A spirit? A god? A thing? It didn't look like some-

An *Apollo 11* photograph of the heavily cratered lunar highlands. Courtesy NASA.

thing big far away, but more like something small nearby—some-thing the size of a plate, maybe, hanging in the sky a little above our heads. Ancient Greek philosophers debated the proposition "that the Moon is exactly as large as it looks" (betraying a hopeless confusion between linear and angular size). *Walking* on the Moon would have seemed a screwball idea; it made more sense to imag-ine somehow climbing up into the sky on a ladder or on the back of a giant bird, grabbing the Moon, and bringing it down to Earth. Nobody ever succeeded, although there were myths aplenty about heroes who had tried.

Not until a few centuries ago did the idea of the Moon as a *place,* a quarter-million miles away, gain wide currency. And in that brief flicker of time, we've gone from the earliest steps in under-standing the Moon's nature to walking and joy-riding on its surface. We calculated how objects move in space; liquefied oxygen from the air; invented big rockets, telemetry, reliable electronics, inertial guidance, and much else. Then we sailed out into the sky.

I was lucky enough to be involved in the *Apollo* program, but I don't blame people who think the whole thing was faked in a Hollywood movie studio. In the late Roman Empire, pagan philosophers had attacked Christian doctrine on the ascension to Heaven of the body of Christ and on the promised bodily resur-rection of the dead—because the force of gravity pulls down all "earthly bodies." St. Augustine rejoined: "If human skill can by some contrivance fabricate vessels that float, out of metals which sink . . . how much more credible is it that God, by some hidden mode of operation, should even more certainly effect that these earthly masses be emancipated" from the chains that bind them to Earth? That *humans* should one day discover such a "mode of op-eration" was beyond imagining. Fifteen hundred years later, we emancipated ourselves.

The achievement elicited an amalgam of awe and concern. Some remembered the story of the Tower of Babel. Some, ortho-dox Moslems among them, felt setting foot on the Moon's surface to be impudence and sacrilege. Many greeted it as a turning point in history.

The Moon is no longer unattainable. A dozen humans, all Americans, have made those odd bounding motions they called "moonwalks" on the crunchy, cratered, ancient gray lava—beginning on that July day in 1969. But since 1972, no one from any nation has ventured back. Indeed, none of us has gone *anywhere* since the glory days of *Apollo* except into low Earth orbit—like a toddler who takes a few tentative steps outward and then, breathless, retreats to the safety of his mother's skirts.

Once upon a time, we soared into the Solar System. For a few years. Then we hurried back. Why? What happened? What was *Apollo* really about?

The scope and audacity of John Kennedy's May 25, 1961, message to a joint session of Congress on "Urgent National Needs"—the speech that launched the *Apollo* program—dazzled me. We would use rockets not yet designed and alloys not yet conceived, navigation and docking schemes not yet devised, in order to send a man to an unknown world—a world not yet explored, not even in a preliminary way, not even by robots—and we would bring him safely back, and we would do it before the decade was over. This confident pronouncement was made before any American had even achieved Earth orbit.

As a newly minted Ph.D., I actually thought all this had something centrally to do with science. But the President did not talk about discovering the origin of the Moon, or even about bringing samples of it back for study. All he seemed to be interested in was sending someone there and bringing him home. It was a kind of *gesture*. Kennedy's science advisor, Jerome Wiesner, later told me he had made a deal with the President: If Kennedy would not claim that *Apollo* was about science, then he, Wiesner, would support it. So if not science, what?

The *Apollo* program is really about politics, others told me. This sounded more promising. Nonaligned nations would be tempted to drift toward the Soviet Union if it was ahead in space exploration, if the United States showed insufficient "national vigor." I didn't follow. Here was the United States, ahead of the Soviet Union in virtually every area of technology—the world's economic, military, and, on occasion, even moral leader—and Indonesia would go Communist because Yuri Gagarin beat John Glenn to Earth orbit? What's so special about space technology? Suddenly I understood.

In the 1969 *Apollo 10* mission, the Command Service Module is seen over the lunar surface. Courtesy NASA.

Sending people to orbit the Earth or robots to orbit the Sun requires rockets—big, reliable, powerful rockets. Those same rockets can be used for nuclear war. The same technology that transports a man to the Moon can carry nuclear warheads halfway around the world. The same technology that puts an astronomer and a telescope in Earth orbit can also put up a laser "battle station." Even back then, there was fanciful talk in military circles, East and West, about space as the new "high ground," about the nation that "controlled" space "controlling" the Earth. Of course strategic rockets were already being tested on Earth. But heaving a ballistic missile with a dummy warhead into a target zone in the middle of the Pacific Ocean doesn't buy much glory. Sending people into space captures the attention and imagination of the world.

You wouldn't spend the money to launch astronauts for this reason alone, but of all the ways of demonstrating rocket potency,

The *Apollo 11* Lunar Module rising off the lunar surface. Courtesy NASA.

this one works best. It was a rite of national manhood; the shape of the boosters made this point readily understood without anyone actually having to explain it. The communication seemed to be transmitted from unconscious mind to unconscious mind without the higher mental faculties catching a whiff of what was going on.

My colleagues today—struggling for every space science dollar—may have forgotten how easy it was to get money for "space" in the glory days of *Apollo* and just before. Of many examples, consider this exchange before the Defense Appropriations Subcommittee of the House of Representatives in 1958, only a few months after *Sputnik 1*. Air Force Assistant Secretary Richard E. Horner is testifying; his interlocutor is Rep. Daniel J. Flood (Democrat of Pennsylvania):

> HORNER: [W]hy is it desirable from the military point of view to have a man on the moon? Partly, from the classic point of view, because it is there. Partly because we might be afraid that the U.S.S.R. might get one there first and realize advantages which we had not anticipated existed there . . .

Earthrise, *Apollo 15.*
Courtesy NASA.

FLOOD: [I]f we gave you all the money you said was necessary, regardless of how much it was, can you in the Air Force hit the moon with something, anything, before Christmas?

HORNER: I feel sure we can. There is always a certain amount of risk in this kind of undertaking, but we feel that we can do that; yes, sir.

FLOOD: Have you asked anybody in the Air Force or the Department of Defense to give you enough money, hardware, and people, starting at midnight tonight, to chip a piece out of that ball of green cheese for a Christmas present to Uncle Sam? Have you asked for that?

HORNER: We have submitted such a program to the Office of the Secretary of Defense. It is currently under consideration.

FLOOD: I am for giving it to them as of this minute, Mr. Chairman, with our supplemental, without waiting for somebody downtown to make up his mind to ask for it. If this man means what he says and if he knows what he is talking about—and I think he does—then this committee should not wait five minutes more today. We should give him all the money and all the hardware and all the people he wants, regardless of what anybody else says or wants, and tell him to go up on top of some hill and do it without any question.

When President Kennedy formulated the *Apollo* program, the Defense Department had a slew of space projects under development—ways of carrying military personnel up into space, means of conveying them around the Earth, robot weapons on orbiting platforms intended to shoot down satellites and ballistic missiles of other nations. *Apollo* supplanted these programs. They never reached operational status. A case can be made then that *Apollo* served another purpose—to move the U.S.-Soviet space competition from a military to a civilian arena. There are some who believe that Kennedy intended *Apollo* as a substitute for an arms race in space. Maybe.

For me, the most ironic token of that moment in history is the plaque signed by President Richard M. Nixon that *Apollo 11* took to the Moon. It reads: "We came in peace for all mankind." As the United States was dropping $7\frac{1}{2}$ megatons of conventional explosives on small nations in Southeast Asia, we congratulated ourselves on our humanity: We would harm no one on a lifeless rock. That plaque is there still, attached to the base of the *Apollo 11* Lunar Module, on the airless desolation of the Sea of Tranquility. If no one disturbs it, it will still be readable a million years from now.

Six more missions followed *Apollo 11,* all but one of which successfully landed on the lunar surface. *Apollo 17* was the first to carry a scientist. As soon as he got there, the program was can-

A lunar landscape. NASA imagery; panorama digitally processed by Artis Planetarium, Amsterdam.

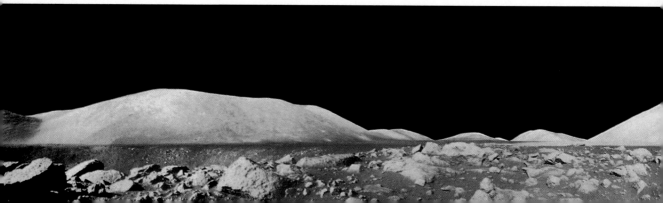

Galileo spacecraft mosaic of the Earth's Moon. False color is used to bring out mineral deposits. Green and yellow indicate a greater abundance of iron and magnesium. Photomosaic courtesy USGS/NASA.

celed. The first scientist and the last human to land on the Moon were the same person. The program had already served its purpose that July night in 1969. The half-dozen subsequent missions were just momentum.

Apollo was not mainly about science. It was not even mainly about space. *Apollo* was about ideological confrontation and nuclear war—often described by such euphemisms as world "leadership" and national "prestige." Nevertheless, good space science was

The Earth and the Moon to scale. The Earth is almost four times the diameter of the Moon and 81 times more massive. The Earth—seen here are the Americas from New England to Patagonia—on average reflects back to space about four times as much light as the dusky Moon. Image courtesy USGS.

done. We now know much more about the composition, age, and history of the Moon and the origin of the lunar landforms. We have made progress in understanding where the Moon came from. Some of us have used lunar cratering statistics to better understand the Earth at the time of the origin of life. But more important than any of this, *Apollo* provided an aegis, an umbrella under which brilliantly engineered robot spacecraft were dispatched throughout the Solar System, making that preliminary reconnaissance of dozens of worlds. The offspring of *Apollo* have now reached the planetary frontiers.

If not for *Apollo*—and, therefore, if not for the political purpose it served—I doubt whether the historic American expeditions of exploration and discovery throughout the Solar System would have occurred. The *Mariners, Vikings, Pioneers, Voyagers,* and *Galileo* are among the gifts of *Apollo. Magellan* and *Cassini* are more distant descendants. Something similar is true for the pioneering Soviet efforts in Solar System exploration, including the first soft landings of robot spacecraft—*Luna 9, Mars 3, Venera 8*—on other worlds.

Apollo conveyed a confidence, energy, and breadth of vision that did capture the imagination of the world. That too was part of its purpose. It inspired an optimism about technology, an enthusiasm for the future. If we could fly to the Moon, as so many have asked, what else were we capable of? Even those who opposed the policies and actions of the United States—even those who thought the worst of us—acknowledged the genius and heroism of the *Apollo* program. With *Apollo,* the United States touched greatness.

When you pack your bags for a big trip, you never know what's in store for you. The *Apollo* astronauts on their way to and from the Moon photographed their home planet. It was a natural thing to do, but it had consequences that few foresaw. For the first time, the inhabitants of Earth could see their world from above— the whole Earth, the Earth in color, the Earth as an exquisite spinning white and blue ball set against the vast darkness of space. Those images helped awaken our slumbering planetary consciousness. They provide incontestable evidence that we all share the same vulnerable planet. They remind us of what is important and what is not. They were the harbingers of *Voyager's* pale blue dot.

We may have found that perspective just in time, just as our technology threatens the habitability of our world. Whatever the reason we first mustered the *Apollo* program, however mired it was in Cold War nationalism and the instruments of death, the inescapable recognition of the unity and fragility of the Earth is its clear and luminous dividend, the unexpected final gift of *Apollo.* What began in deadly competition has helped us to see that global cooperation is the essential precondition for our survival.

Travel is broadening.

It's time to hit the road again.

EXPLORING OTHER WORLDS AND PROTECTING THIS ONE

The planets, in their various stages of development, are subjected to the same formative forces that operate on our earth, and have, therefore, the same geologic formation, and probably life, of our own past, and perhaps future; but, further than this, these forces are acting, in some cases, under totally different conditions from those under which they operate on the earth, and hence must evolve forms different from those ever known to man. The value of such material as this to the comparative sciences is too obvious to need discussion.

ROBERT H. GODDARD, NOTEBOOK (1907)

For the first time in my life, I saw the horizon as a curved line. It was accentuated by a thin seam of dark blue light—our atmosphere. Obviously, this was not the "ocean" of air I had been told it was so many times in my life. I was terrified by its fragile appearance.

—ULF MERBOLD, GERMAN SPACE SHUTTLE ASTRONAUT (1988)

OPPOSITE: A human being orbiting the Earth views the home planet with its "thin seam of dark blue light": Astronaut Bruce McCandless in his Manned Maneuvering Unit (MMU) in February 1984. Photograph taken from the space shuttle *Challenger*. Courtesy Johnson Space Center/NASA.

When you look down at the Earth from orbital altitudes, you see a lovely, fragile world embedded in black vacuum. But peering at a piece of the Earth through a spacecraft porthole is nothing like the joy of seeing it entire against the backdrop of black, or—better—sweeping across your field of view as you float in space unencumbered by a spacecraft. The first human to have this experience was Alexei Leonov, who on March 18, 1965, left *Voskhod 2* in the original space "walk": "I looked down at the Earth," he recalls, "and the first thought that crossed my mind was 'The world *is* round, after all.' In one glance I could see from Gibraltar to the Caspian Sea . . . I felt like a bird—with wings, and able to fly."

When you view the Earth from farther away, as the *Apollo* astronauts did, it shrinks in apparent size, until nothing but a little geography remains. You're struck by how self-contained it is. An occasional hydrogen atom leaves; a pitter-patter of cometary dust arrives. Sunlight, generated in the immense, silent thermonuclear engine deep in the solar interior, pours out of the Sun in all directions, and the Earth intercepts enough of it to provide a little illumination and enough heat for our modest purposes. Apart from that, this small world is on its own.

From the surface of the Moon you can see it, perhaps as a crescent, even its continents now indistinct. And from the vantage point of the outermost planet it is a mere point of pale light.

From Earth orbit, you are struck by the tender blue arc of the horizon—the Earth's thin atmosphere seen tangentially. You can understand why there is no longer such a thing as a local environmental problem. Molecules are stupid. Industrial poisons, greenhouse gases, and substances that attack the protective ozone layer, because of their abysmal ignorance, do not respect borders. They are oblivious of the notion of national sovereignty. And so, due to the almost mythic powers of our technology (and the prevalence of short-term thinking), we are beginning—on continental and on planetary scales—to pose a danger to ourselves. Plainly, if these problems are to be solved, it will require many nations acting in concert over many years.

I'm struck again by the irony that spaceflight—conceived in

Earthrise over the Moon. At this resolution, even the continents of Earth are invisible. *Apollo 14* photograph, courtesy NASA.

the caldron of nationalist rivalries and hatreds—brings with it a stunning transnational vision. You spend even a little time contemplating the Earth from orbit and the most deeply engrained nationalisms begin to erode. They seem the squabbles of mites on a plum.

If we're stuck on one world, we're limited to a single case; we don't know what else is possible. Then—like an art fancier familiar only with Fayoum tomb paintings, a dentist who knows only molars, a philosopher trained merely in Neo-Platonism, a linguist who has studied only Chinese, or a physicist whose knowledge of gravity is restricted to falling bodies on Earth—our perspective is foreshortened, our insights narrow, our predictive abilities circumscribed. By contrast, when we explore other worlds, what once

False-color radar imagery of Mt. Pinatubo in the Philippines. The principal volcanic crater or caldera produced in the great June 1991 explosion can be seen at the boundary of orange/brown and lighter colors. The dark drainage features are mud flows, a continuing hazard in heavy rains. Fine droplets of sulfuric acid punched into the stratosphere by the Pinatubo explosion had nearly worldwide effects, temporarily adding to ozone layer depletion and slowing the warming trend attributed to an increasing greenhouse effect. The image was acquired by the spaceborne imaging radar aboard the space shuttle *Endeavor* on orbit 78, April 13, 1994. Courtesy JPL/NASA.

seemed the only way a planet could be turns out to be somewhere in the middle range of a vast spectrum of possibilities. When we look at those other worlds, we begin to understand what happens when we have too much of one thing or too little of another. We learn how a planet can go wrong. We gain a new understanding, foreseen by the spaceflight pioneer Robert Goddard, called comparative planetology.

The exploration of other worlds has opened our eyes in the study of volcanos, earthquakes, and weather. It may one day have profound implications for biology, because all life on Earth is built on a common biochemical master plan. The discovery of a single extraterrestrial organism—even something as humble as a bacterium—would revolutionize our understanding of living things.

But the connection between exploring other worlds and protecting this one is most evident in the study of Earth's climate and the burgeoning threat to that climate that our technology now poses. Other worlds provide vital insights about what dumb things not to do on Earth.

Three potential environmental catastrophes—all operating on a global scale—have recently been uncovered: ozone layer depletion, greenhouse warming, and nuclear winter. All three discoveries, it turns out, have strong ties to the exploration of the planets:

(1) It was disturbing to find that an inert material with all sorts of practical applications—it serves as the working fluid in refrigerators and air conditioners, as aerosol propellant for deodorants and other products, as lightweight foamy packaging for fast foods, and as a cleaning agent in microelectronics, to name only a few—can pose a danger to life on Earth. Who would have figured?

The molecules in question are called chlorofluorocarbons (CFCs). Chemically, they're extremely inert, which means they're invulnerable—until they find themselves up in the ozone layer, where they're broken apart by ultraviolet light from the Sun. The chlorine atoms thus liberated attack and break down the protective ozone, letting more ultraviolet light reach the ground. This increased ultraviolet intensity ushers in a ghastly procession of potential consequences involving not just skin cancer and cataracts, but weakening of the human immune system and, most dangerous of all, possible harm to agriculture and to photosynthetic organisms at the base of the food chain on which most life on Earth depends.

Who discovered that CFCs posed a threat to the ozone layer? Was it the principal manufacturer, the DuPont Corporation, exercising corporate responsibility? Was it the Environmental Protection Agency protecting us? Was it the Department of Defense defending us? No, it was two ivory-tower, white-coated university scientists working on something else—Sherwood Rowland and Mario Molina of the University of California, Irvine. Not even an Ivy League university. No one instructed them to look for dangers

Ultraviolet-fried soil on Mars. At left, the *Viking* sample arm pushes a vesicular rock out of the way and digs. When the soil sample is retrieved (right) and analyzed inside the spacecraft, not even trace amounts of organic matter can be detected. Unlike Earth, Mars has no ozone shield. *Viking* images, courtesy JPL/NASA.

to the environment. They were pursuing fundamental research. They were scientists following their own interests. Their names should be known to every schoolchild.

In their original calculations, Rowland and Molina used rate constants of chemical reactions involving chlorine and other halogens that had been measured in part with NASA support. Why NASA? Because Venus has chlorine and fluorine molecules in its atmosphere, and planetary aeronomers had wanted to understand what's happening there.

Confirming theoretical work on the role of CFCs in ozone depletion was soon done by a group led by Michael McElroy at Harvard. How is it they had all these branching networks of halogen chemical kinetics in their computer ready to go? Because they were working on the chlorine and fluorine chemistry of the atmosphere of Venus. Venus helped make and helped confirm the discovery that the Earth's ozone layer is in danger. An entirely unexpected connection was found between the atmospheric photochemistries of the two planets. A result of importance to everyone on Earth emerged from what might well have seemed the most blue-sky, abstract, impractical kind of work, understanding the chemistry of minor constituents in the upper atmosphere of another world.

There's also a Mars connection. With *Viking* we found the surface of Mars to be apparently lifeless and remarkably deficient even in simple organic molecules. But simple organic molecules *ought* to be there, because of the impact of organic-rich meteorites from the nearby asteroid belt. This deficiency is widely attributed to the lack of ozone on Mars. The *Viking* microbiology experiments found that organic matter carried from Earth to Mars and sprinkled on Martian surface dust is quickly oxidized and destroyed. The materials in the dust that do the destruction are molecules something like hydrogen peroxide—which we use as an antiseptic because it kills microbes by oxidizing them. Ultraviolet light from the Sun strikes the surface of Mars unimpeded by an ozone layer; if any organic matter *were* there, it would be quickly destroyed by the ultraviolet light itself and its oxidation products. Thus part of the reason the topmost layers of Martian soil are antiseptic is that Mars has an ozone hole of planetary dimensions—by itself a useful cautionary tale for us, who are busily thinning and puncturing *our* ozone layer.

(2) Global warming is predicted to follow from the increasing greenhouse effect caused largely by carbon dioxide generated in the burning of fossil fuels—but also from the buildup of other infrared-absorbing gases (oxides of nitrogen, methane, those same CFCs, and other molecules).

Suppose that we have a three-dimensional general circulation computer model of the Earth's climate. Its programmers claim it's able to predict what the Earth will be like if there's more of one atmospheric constituent or less of another. The model does very well at "predicting" the present climate. But there is a nagging worry: The model has been "tuned" so it will come out right— that is, certain adjustable parameters are chosen, not from first principles of physics, but to get the right answer. This is not exactly cheating, but if we apply the same computer model to rather different climatic regimes—deep global warming, for instance—the tuning might then be inappropriate. The model might be valid for today's climate, but not extrapolatable to others.

One way to test this program is to apply it to the very different climates of other planets. Can it predict the structure of the atmosphere on Mars and the climate there? The weather? What about Venus? If it were to fail these test cases, we would be right in mistrusting it when it makes predictions for our own planet. In fact, climate models now in use do very well in predicting from first principles of physics the climates on Venus and Mars.

On Earth, huge upwellings of molten lava are known and attributed to superplumes convecting up from the deep mantle and generating vast plateaus of frozen basalt. A spectacular example occurred about a hundred million years ago, and added perhaps ten times the present carbon dioxide content to the atmosphere, inducing substantial global warming. These plumes, it is thought, occur episodically throughout Earth's history. Similar mantle upwelling seem to have occurred on Mars and Venus. There are sound practical reasons for us to want to understand how a major change to the Earth's surface and climate could suddenly arrive unannounced from hundreds of kilometers beneath our feet.

Some of the most important recent work on global warming has been done by James Hansen and his colleagues at the Goddard Institute for Space Sciences, a NASA facility in New York City. Hansen developed one of the major computer climate models and employed it to predict what will happen to our climate as the greenhouse gases continue to build up. He has been in the forefront of testing these models against ancient climates of the Earth. (During the last ice ages, it is of interest to note, more carbon dioxide and methane are strikingly correlated with higher temperatures.) Hansen collected a wide range of weather data from this century and last, to see what actually happened to the global temperature, and then compared it to the computer model's predictions of what *should* have happened. The two agree to within the errors of measurement and calculation, respectively. He coura-

The scorching surface of Venus, as first seen directly by the *Venera* spacecraft, is due to a massive carbon dioxide greenhouse effect. Courtesy Vernadsky Institute, Moscow.

geously testified before Congress in the face of a politically generated order from the White House Office of Management and Budget (this was in the Reagan years) to exaggerate the uncertainties and minimize the dangers. His calculation on the explosion of the Philippine volcano Mt. Pinatubo and his prediction of the resulting temporary decline in the Earth's temperature (about half a degree Celsius) were right on the money. He has been a force in convincing governments worldwide that global warming is something to be taken seriously.

How did Hansen get interested in the greenhouse effect in the first place? His doctoral thesis (at the University of Iowa in 1967) was about Venus. He agreed that the high radio brightness of Venus is due to a very hot surface, agreed that greenhouse gases keep the heat in, but proposed that heat from the interior rather than sunlight was the principal energy source. The *Pioneer 12* mission to Venus in 1978 dropped entry probes into the atmosphere; they showed directly that the ordinary greenhouse effect—the surface heated by the Sun and the heat retained by the blanket of

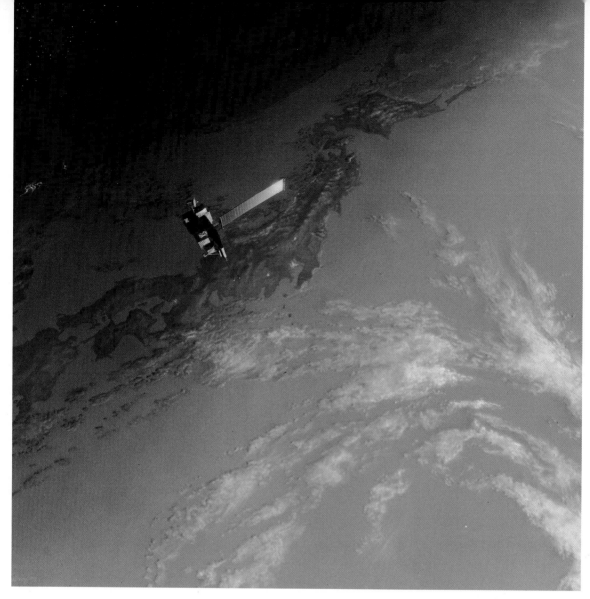

A future Earth observation satellite over Japan. Painting by Pat Rawlings, copyright Pat Rawlings ©1994.

air—was the operative cause. But it's Venus that got Hansen thinking about the greenhouse effect.

Radio astronomers, you note, find Venus to be an intense source of radio waves. Other explanations of the radio emission fail. You conclude that the surface must be ridiculously hot. You try to understand where the high temperatures come from and are led inexorably to one or another kind of greenhouse effect. Decades later you find that this training has prepared you to understand and help predict an unexpected threat to our global civilization. I know many other instances where scientists who first tried to puzzle out the atmospheres of other worlds are making important and highly practical discoveries about this one. The other planets are a superb training ground for students of the Earth. They require both breadth and depth of knowledge, and they challenge the imagination.

Those who are skeptical about carbon dioxide greenhouse warming might profitably note the massive greenhouse effect on Venus. No one proposes that Venus's greenhouse effect derives from imprudent Venusians who burned too much coal, drove fuel-inefficient autos, and cut down their forests. My point is different. The climatological history of our planetary neighbor, an otherwise Earthlike planet on which the surface became hot enough to melt tin or lead, is worth considering—especially by those who say that the increasing greenhouse effect on Earth will be self-correcting, that we don't really have to worry about it, or (you can see this in the publications of some groups that call themselves conservative) that the greenhouse effect itself is a "hoax."

(3) Nuclear winter is the predicted darkening and cooling of the Earth—mainly from fine smoke particles injected into the atmosphere from the burning of cities and petroleum facilities—that is predicted to follow a global thermonuclear war. A vigorous scientific debate ensued on just how serious nuclear winter might be. The various opinions have now converged. All three-dimensional general circulation computer models predict that the global temperatures resulting from a worldwide thermonuclear war would be colder than those in the Pleistocene ice ages. The implications for our planetary civilization—especially through the collapse of agriculture—are very dire. It is a consequence of nuclear war that was somehow overlooked by the civil and military authorities of the United States, the Soviet Union, Britain, France, and China when they decided to accumulate well over 60,000 nuclear weapons. Although it's hard to be certain about such things, a case can be made that nuclear winter played a constructive role (there were other causes, of course) in convincing the nuclear-armed nations, especially the Soviet Union, of the futility of nuclear war.

Nuclear winter was first calculated and named in 1982/83 by a group of five scientists, to which I'm proud to belong. This team was given the acronym TTAPS (for Richard P. Turco, Owen B.

Toon, Thomas Ackerman, James Pollack, and myself). Of the five TTAPS scientists, two were planetary scientists, and the other three had published many papers in planetary science. The earliest intimation of nuclear winter came during that same *Mariner 9* mission to Mars, when there was a global dust storm and we were unable to see the surface of the planet; the infrared spectrometer on the spacecraft found the high atmosphere to be warmer and the surface colder than they ought to have been. Jim Pollack and I sat down and tried to calculate how that could come about. Over the subsequent twelve years, this line of inquiry led from dust storms on Mars to volcanic aerosols on Earth to the possible extinction of the dinosaurs by impact dust to nuclear winter. You never know where science will take you.

PLANETARY SCIENCE fosters a broad interdisciplinary point of view that proves enormously helpful in discovering and attempting to defuse these looming environmental catastrophes. When you cut your teeth on other worlds, you gain a perspective about the fragility of planetary environments and about what other, quite different, environments are possible. There may well be potential global catastrophes still to be uncovered. If there are, I bet planetary scientists will play a central role in understanding them.

Of all the fields of mathematics, technology, and science, the one with the greatest international cooperation (as determined by how often the co-authors of research papers hail from two or more countries) is the field called "Earth and space sciences." Studying this world and others, by its very nature, tends to be non-local, non-nationalist, non-chauvinist. Very rarely do people go into these fields *because* they are internationalists. Almost always, they enter for other reasons, and then discover that splendid work, work that complements their own, is being done by researchers in other nations; or that to solve a problem, you need data or a perspective (access to the southern sky, for example) that is unavailable in your country. And once you experience such cooperation —humans from different parts of the planet working in a mutually intelligible scientific language as partners on matters of common concern—it's hard not to imagine it happening on other, nonsci-

"Stewardship," watercolor by Greg Mort.

entific matters. I myself consider this aspect of Earth and space sciences as a healing and unifying force in world politics; but, beneficial or not, it is inescapable.

When I look at the evidence, it seems to me that planetary exploration is of the most practical and urgent utility for us here on Earth. Even if we were not roused by the prospect of exploring other worlds, even if we didn't have a nanogram of adventuresome spirit in us, even if we were only concerned for ourselves and in the narrowest sense, planetary exploration would still constitute a superb investment.

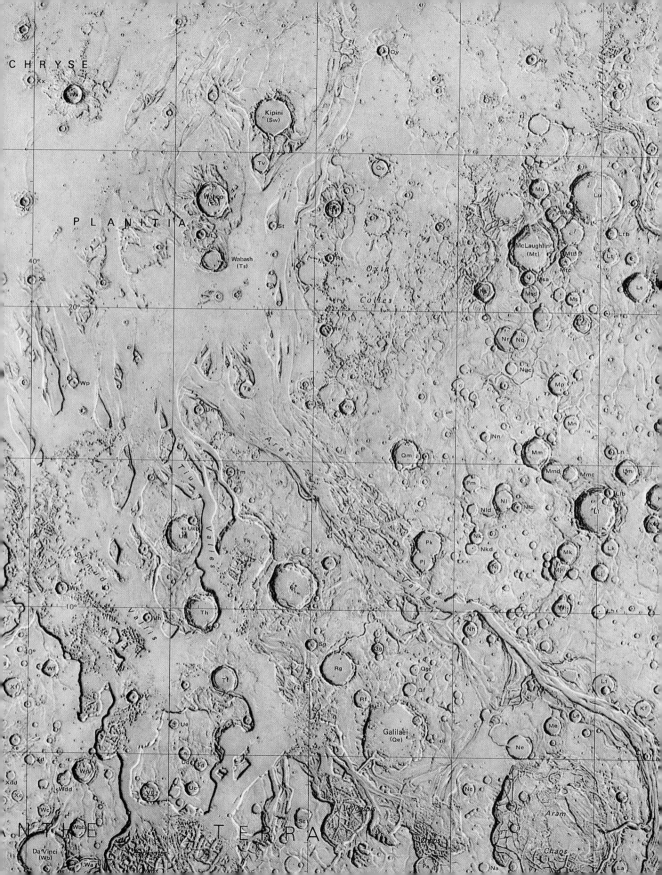

CHRYSE

PLANITIA

40°

20°

THE   TERRA

Kipini
(Sw)

Wahoo
(Tb)

Wabash
(Ts)

McLaughlin
(Mt)

Oxia

Colles

Ares

Tiu

Simud

Valles

Valles

Valles

Ukb

Galilaei
(Qe)

Hydaspis

Chaos

Aram

Chaos

Da Vinci
(Wb)

Da Vinci

THE GATES OF THE WONDER WORLD OPEN

The great floodgates of the wonder-world swung open.

—HERMAN MELVILLE, *MOBY DICK*, CHAPTER 1 (1851)

The Oxia Palus region of Mars, between the Equator and latitude 25° north. The *Viking 1* landing site is in Chryse Planitia at top left. Great floods dissected this terrain billions of years ago. But the sparseness here of tributaries in the river valleys points to water that upwelled from underground, rather than that fell as rain from the skies. *Viking 1* had originally been intended to land in the confluence of these outflow channels, but safety considerations intervened. The reason that the Martian climate is now so far from the inferred warmer and wetter environment of four billion years ago is unknown. Toward lower right is a crater named after Galileo. USGS shaded relief map.

Sometime coming up, perhaps just around the corner, there will be a nation—more likely, a consortium of nations—that will work the next major step in the human venture into space. Perhaps it will be brought about by circumventing bureaucracies and making efficient use of present technologies. Perhaps it will require new technologies, transcending the great blunderbuss chemical rockets. The crews of these ships will set foot on new worlds. The first baby will be born somewhere up there. Early steps toward living off the land will be made. We will be on our way. And the future will remember.

TANTALIZING AND MAJESTIC, Mars is the world next door, the nearest planet on which an astronaut or cosmonaut could safely land. Although it is sometimes as warm as a New England October, Mars is a chilly place, so cold that some of its thin carbon dioxide atmosphere freezes out as dry ice at the winter pole.

It is the nearest planet whose surface we can see with a small telescope. In all the Solar System, it is the planet most like Earth. Apart from flybys, there have been only two fully successful missions to Mars: *Mariner 9* in 1971, and *Vikings 1* and *2* in 1976. They revealed a deep rift valley that would stretch from New York to San Francisco; immense volcanic mountains, the largest of which towers 80,000 feet above the average altitude of the Martian surface, almost three times the height of Mount Everest; an intricate layered structure in and among the polar ices, resembling a pile of discarded poker chips, and probably a record of past climatic change; bright and dark streaks painted down on the surface by windblown dust, providing high-speed wind maps of Mars over the past decades and centuries; vast globe-girdling dust storms; and enigmatic surface features.

Hundreds of sinuous channels and valley networks dating back several billion years can be found, mainly in the cratered southern highlands. They suggest a previous epoch of more benign and Earthlike conditions—very different from what we find beneath the tenuous and frigid atmosphere of our time. Some ancient channels seem to have been carved by rainfall, some by underground sapping and collapse, and some by great floods that gushed up out of the ground. Rivers were pouring into and filling great thousand-kilometer-diameter impact basins that today are dry as dust. Waterfalls dwarfing any on Earth today cascaded into the lakes of ancient Mars. Vast oceans, hundreds of meters, perhaps even a kilometer, deep may have gently lapped shorelines barely discernible today. *That* would have been a world to explore. We are four billion years late.*

On Earth in just the same period, the first microorganisms arose and evolved. Life on Earth is intimately connected, for the most basic chemical reasons, with liquid water. We humans are ourselves *made* of some three-quarters water. The same sorts of organic molecules that fell out of the sky and were generated in the air and seas of ancient Earth, should also have accumulated on ancient Mars. Is it plausible that life quickly came to be in the waters of early Earth, but was somehow restrained and inhibited in the waters of early Mars? Or might the Martian seas have been filled with life—floating, spawning, evolving? What strange beasts once swum there?

Whatever the dramas of those distant times, it all started to go wrong around 3.8 billion years ago. We can see that the erosion of ancient craters dramatically began to slow about then. As the atmosphere thinned, as the rivers flowed no more, as the oceans began to dry, as the temperatures plummeted, life would have retreated to the few remaining congenial habitats, perhaps huddling at the bottom of ice-covered lakes, until it too vanished and the dead bodies and fossil remains of exotic organisms—built, it might be, on principles very different from life on Earth—were deep-frozen, awaiting the explorers who might in some distant future arrive on Mars.

* Although in a few places, such as the slopes of the elevation called Alba Patera, there are multibranched valley networks that by comparison are very young. Somehow, even in the most recent billion years, liquid water seems to have flowed here and there, from time to time, through the deserts of Mars.

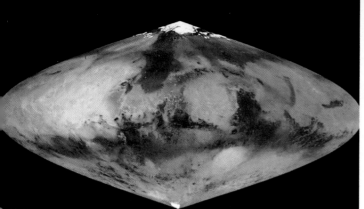

Equal-area projection of all of Mars in a *Viking* photomosaic. The large northern and the small southern caps wax and wane with the seasons. The faintly yellowish, circular feature in the south is Hellas, a giant dust-filled impact basin that billions of years ago may have been a lake. The resolution in this picture is roughly comparable to that of the best telescopic observations from the surface of the Earth. It is perhaps possible to see why some of the early visual observers had an impression of "canals." Courtesy USGS/NASA.

The great 5,000-kilometer-long rift valley named Vallis Marineris, commemorating the *Mariner 9* spacecraft that first revealed the modern Mars. West of it is seen one of the shield volcanos on the Elysium Plateau. *Viking* photomosaic, courtesy USGS/NASA.

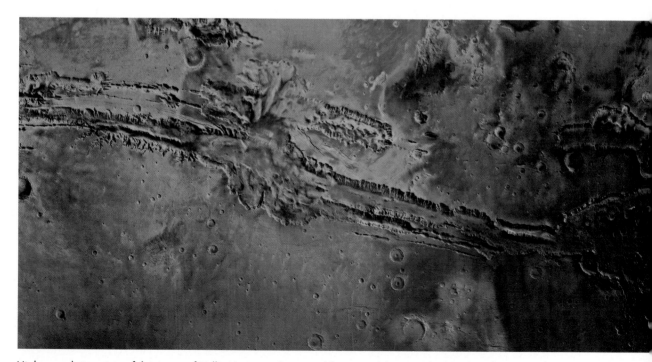

Higher-resolution view of the center of Vallis Marineris. Its vertical faces are kilometers high. *Viking* photomosaic, courtesy USGS/NASA.

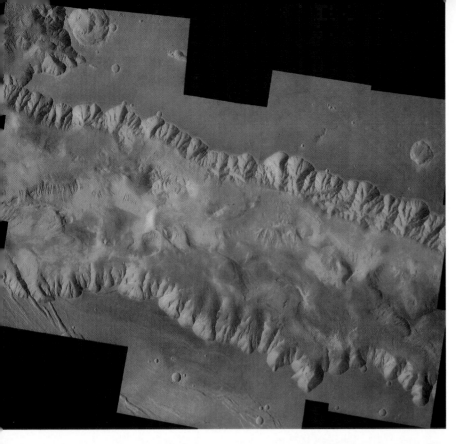

Details in the interior of Vallis Marineris. The valley floor is filled with debris that has slumped down from the cliff faces in enormous avalanches. *Viking* photomosaics, courtesy USGS/NASA.

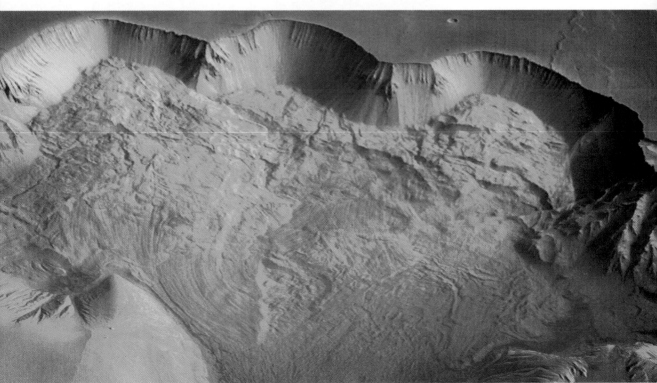

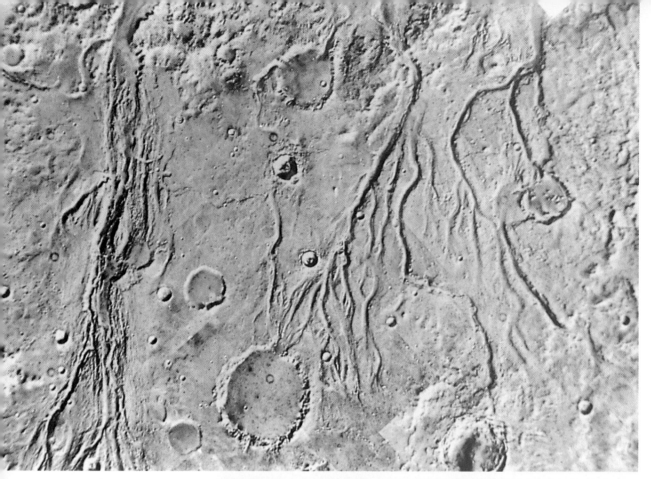

Photomosaic of a rich river valley
network. *Mariner 9* photomosaic,
courtesy JPL/NASA.

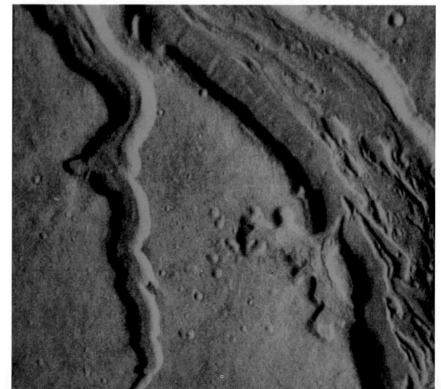

Interior of a Martian river valley
as seen by *Viking*. Such locales
are prime sites for future roving
vehicles. Courtesy JPL/NASA.

METEORITES ARE FRAGMENTS OF OTHER WORLDS recovered on Earth. Most originate in collisions among the numerous asteroids that orbit the Sun between the orbits of Mars and Jupiter. But a few are generated when a large meteorite impacts a planet or asteroid at high speed, gouges out a crater, and propels the excavated surface material into space. A very small fraction of the ejected rocks, millions of years later, may intercept another world.

In the wastelands of Antarctica, the ice is here and there dotted with meteorites, preserved by the low temperatures and until recently undisturbed by humans. A few of them, called SNC (pronounced "snick") meteorites [*] have an aspect about them that at first seemed almost unbelievable: Deep inside their mineral and glassy structures, locked away from the contaminating influence of the Earth's atmosphere, a little gas is trapped. When the gas is analyzed, it turns out to have exactly the same chemical composition and isotopic ratios as the air on Mars. We know about Martian air not just from spectroscopic inference but from direct measurement on the Martian surface by the *Viking* landers. To the surprise of nearly everyone, the SNC meteorites come from Mars.

Originally, they were rocks that had melted and refrozen. Radioactive dating of all the SNC meteorites shows their parent rocks condensed out of lava between 180 million and 1.3 billion years ago. Then they were driven off the planet by collisions from space. From how long they've been exposed to cosmic rays on their interplanetary journeys between Mars and Earth, we can tell how old they are—how long ago they were ejected from Mars. In this sense, they are between 10 million and 700,000 years old. They sample the most recent 0.1 percent of Martian history.

Some of the minerals they contain show clear evidence of having once been in water, warm liquid water. These hydrothermal minerals reveal that somehow, probably all over Mars, there was recent liquid water. Perhaps it came about when the interior heat melted underground ice. But however it happened, it's natural to wonder if life is not entirely extinct, if somehow it's managed to hang on into our time in transient underground lakes, or even in thin films of water wetting subsurface grains.

---

[*] Short for Shergotty-Nakhla-Chassigny. You can see why the acronym is used.

Six of the ten known SNC
meteorites, as if on their way
to Earth from Mars in this
photomontage. The oxygen and
hydrogen bound to the minerals
in these meteorites have isotopic
compositions characteristic of
the Martian atmosphere. Courtesy
Johnson Space Center/NASA.

The geochemists Everett Gibson and Hal Karlsson of NASA's
Johnson Space Flight Center have extracted a single drop of water
from one of the SNC meteorites. The isotopic ratios of the oxygen and hydrogen atoms that it contains are literally unearthly. I
look on this water from another world as an encouragement for
future explorers and settlers.

Imagine what we might find if a large number of samples, including never melted soil and rocks, were returned to Earth from
Martian locales selected for their scientific interest. We are very
close to being able to accomplish this with small roving robot vehicles.

The transportation of subsurface material from world to

world raises a tantalizing question: Four billion years ago there were two neighboring planets, both warm, both wet. Impacts from space, in the final stages of the accretion of these planets, were occurring at a much higher rate than today. Samples from each world were being flung out into space. We are sure there was life on at least one of them in this period. We know that a fraction of the ejected debris stays cool throughout the processes of impact, ejection, and interception by another world. So could some of the early organisms on Earth have been safely transplanted to Mars four billion years ago, initiating life on that planet? Or, even more speculative, could life on Earth have arisen by such a transfer from Mars? Might the two planets have regularly exchanged life-forms for hundreds of millions of years? The notion might be testable. If we were to discover life on Mars and found it very similar to life on Earth—and if, as well, we were sure it wasn't microbial contamination that we ourselves had introduced in the course of our explorations—the proposition that life was long ago transferred across interplanetary space would have to be taken seriously.

IT WAS ONCE THOUGHT that life is abundant on Mars. Even the dour and skeptical astronomer Simon Newcomb (in his *Astronomy for Everybody,* which went through many editions in the early decades of this century and was the astronomy text of my childhood) concluded, "There appears to be life on the planet Mars. A few years ago this statement was commonly regarded as fantastic. Now it is commonly accepted." Not "intelligent human life," he was quick to add, but green plants. However, we have now been to Mars and looked for plants—as well as animals, microbes, and intelligent beings. Even if the other forms were absent, we might have imagined, as in Earth's deserts today, and as on Earth for almost all its history, abundant microbial life.

The "life detection" experiments on *Viking* were designed to detect only a certain subset of conceivable biologies; they were biased to find the kind of life about which we know. It would have been foolish to send instruments that could not even detect life on Earth. They were exquisitely sensitive, able to find microbes in the most unpromising, arid deserts and wastelands on Earth.

One experiment measured the gases exchanged between

A drop of Martian water, extracted from a SNC meteorite. Courtesy Johnson Space Center/NASA.

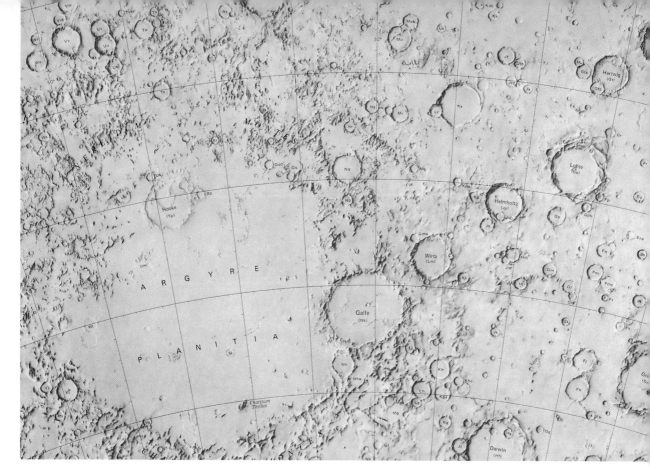

A giant impact basin on Mars: Argyre Planitia in the Southern Hemisphere. There is some evidence that billions of years ago it was filled with water. USGS shaded relief map.

Martian soil and the Martian atmosphere in the presence of organic matter from Earth. A second brought a wide variety of organic foodstuffs marked by a radioactive tracer to see if there were bugs in the Martian soil who ate the food and oxidized it to radioactive carbon dioxide. A third experiment introduced radioactive carbon dioxide (and carbon monoxide) to the Martian soil to see if any of it was taken up by Martian microbes. To the initial astonishment of, I think, all the scientists involved, each of the three experiments gave what at first seemed to be positive results. Gases were exchanged; organic matter was oxidized; carbon dioxide was incorporated into the soil.

But there are reasons for caution. These provocative results are not generally thought to be good evidence for life on Mars: The putative metabolic processes of Martian microbes occurred under a very wide range of conditions inside the *Viking* landers—wet (with liquid water brought from Earth) and dry, light and dark, cold (only a little above freezing) to hot (almost the normal boiling point of water). Many microbiologists deem it unlikely that

Martian microbes would be so capable under such varied conditions. Another strong inducement to skepticism is that a fourth experiment, to look for organic chemicals in the Martian soil, gave uniformly negative results despite its sensitivity. We expect life on Mars, like life on Earth, to be organized around carbon-based molecules. To find no such molecules at all was daunting for optimists among the exobiologists.

The apparently positive results of the life detection experiments is now generally attributed to chemicals that oxidize the soil, deriving ultimately from ultraviolet sunlight (as discussed in the previous chapter). There is still a handful of *Viking* scientists who wonder if there might be extremely tough and competent organisms very thinly spread over the Martian soil—so their organic chemistry could not be detected, but their metabolic processes could. Such scientists do not deny that ultraviolet-generated oxidants are present in the Martian soil, but stress that no

The Margaritifer Sinus region of Mars, between the Equator and 30° south. The east end of the great Vallis Marineris rift valley is seen at top left. Many small-scale river valleys with tributaries are in this province of Mars. Were they made by rainfall? USGS shaded relief map.

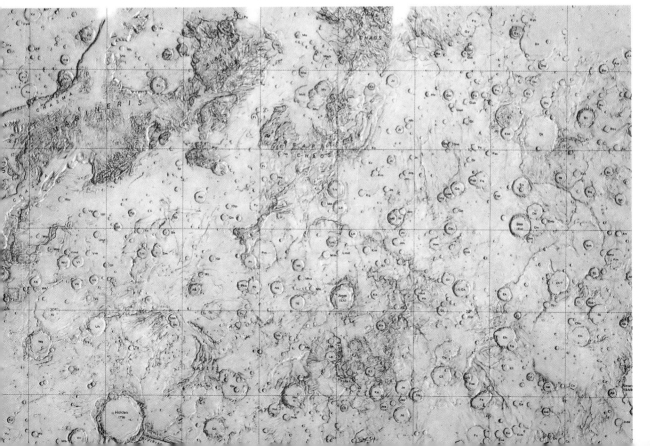

thorough explanation of the *Viking* life detection results from oxidants alone has been forthcoming. Tentative claims have been made of organic matter in SNC meteorites, but they seem instead to be contaminants that have entered the meteorite after its arrival on our world. So far, there are no claims of Martian microbes in these rocks from the sky.

Perhaps because it seems to pander to public interest, NASA and most *Viking* scientists have been very chary about pursuing the biological hypothesis. Even now, much more could be done in going over the old data, in looking with *Viking*-type instruments at Antarctic and other soils that have few microbes in them, in laboratory simulation of the role of oxidants in the Martian soil, and in designing experiments to elucidate these matters—not excluding further searches for life—with future Mars landers.

If indeed no unambiguous signatures of life were determined by a variety of sensitive experiments at two sites 5,000 kilometers apart on a planet marked by global wind transport of fine particles, this is at least suggestive that Mars may be, today at least, a lifeless planet. But if Mars *is* lifeless, we have two planets, of virtually identical age and early conditions, evolving next door to one an-

Viking 1 on Mars. The tall structure at right is the boom supporting the high-gain antenna which returned data to and received commands from Earth. Courtesy JPL/NASA.

Sand dunes at the *Viking 1* site. Courtesy JPL/NASA.

other in the same solar system: Life evolves and proliferates on one, but not the other. Why?

Perhaps the chemical or fossil remains of early Martian life can still be found—subsurface, safely protected from the ultraviolet radiation and its oxidation products that today fry the surface. Perhaps in a rock face exposed by a landslide, or in the banks of an ancient river valley or dry lake bed, or in the polar, laminated terrain, key evidence for life on another planet is waiting.

Despite its absence on the surface of Mars, the planet's two moons, Phobos and Deimos, seem to be rich in complex organic matter dating back to the early history of the Solar System. The Soviet *Phobos 2* spacecraft found evidence of water vapor being outgassed from Phobos, as if it has an icy interior heated by radioactivity. The moons of Mars may have long ago been captured from somewhere in the outer Solar System; conceivably, they are among the nearest available examples of unaltered stuff from the earliest days of the Solar System. Phobos and Deimos are very small, each roughly 10 kilometers across; the gravity they exert is nearly negligible. So it's comparatively easy to rendezvous with

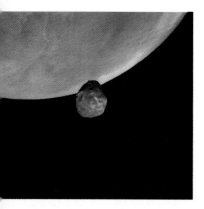

The inner moon Phobos above the Martian surface. Image obtained by the Soviet *Phobos 2* mission; data processed at USGS.

The outer Martian moon Deimos, shown here as a negative to bring out detail. *Viking* image processed at USGS.

them, land on them, examine them, use them as a base of operations to study Mars, and then go home.

Mars calls, a storehouse of scientific information—important in its own right but also for the light it casts on the environment of our own planet. There are mysteries waiting to be resolved about the interior of Mars and its mode of origin, the nature of volcanos on a world without plate tectonics, the sculpting of landforms on a planet with sandstorms undreamt of on Earth, glaciers and polar landforms, the escape of planetary atmospheres, and the capture of moons—to mention a more or less random sampling of scientific puzzles. If Mars once had abundant liquid water and a clement climate, what went wrong? How did an Earthlike world become so parched, frigid, and comparatively airless? Is there something here we should know about our own planet?

We humans have been this way before. The ancient explorers would have understood the call of Mars. But mere scientific exploration does not require a human presence. We can always send smart robots. They are far cheaper, they don't talk back, you can send them to much more dangerous locales, and, with some chance of mission failure always before us, no lives are put at risk.

"HAVE YOU SEEN ME?" the back of the milk carton read. "*Mars Observer*, 6' × 4.5' × 3', 2500 kg. Last heard from on 8/21/93, 627,000 km from Mars."

"*M. O.* call home" was the plaintive message on a banner hung outside the Jet Propulsion Laboratory's Mission Operations Facility in late August 1993. The failure of the United States' *Mars Observer* spacecraft just before it was to insert itself into orbit around Mars was a great disappointment. It was the first postlaunch mission failure of an American lunar or planetary spacecraft in 26 years. Many scientists and engineers had devoted a decade of their professional lives to *M. O.* It was the first U.S. mission to Mars in 17 years—since *Viking*'s two orbiters and two landers in 1976. It was also the first real post–Cold War spacecraft: Russian scientists were on several of the investigator teams, and *Mars Observer* was to act as an essential radio relay link for landers from what was then scheduled to be the Russian *Mars '94* mission, as well as for a daring rover and balloon mission slated for *Mars '96*.

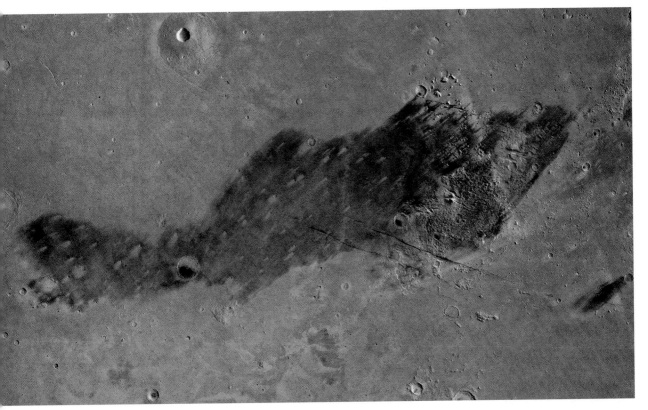

Viking photomosaic in the Elysium Plateau. Fields of bright wind streaks can be seen on the dark ground (as well as a prominent dark streak at bottom right). The high-speed winds that made these streaks veered from the northeast. USGS/NASA.

The scientific instruments aboard *Mars Observer* would have mapped the geochemistry of the planet and prepared the way for future missions, guiding landing site decisions. It might have cast a new light on the massive climate change that seems to have occurred in early Martian history. It would have photographed some of the surface of Mars with detail better than two meters across. Of course, we do not know what wonders *Mars Observer* would have uncovered. But every time we examine a world with new instruments and in vastly improved detail, a dazzling array of discoveries emerges—just as it did when Galileo turned the first telescope toward the heavens and opened the era of modern astronomy.

According to the Commission of Inquiry, the cause of the failure was probably a rupture of the fuel tank during pressurization, gases and liquids sputtering out, and the wounded spacecraft spinning wildly out of control. Perhaps it was avoidable. Perhaps it was an unlucky accident. But to keep this matter in perspective, let's consider the full range of missions to the Moon and the planets attempted by the United States and the former Soviet Union:

In the beginning, our track records were poor. Space vehicles blew up at launch, missed their targets, or failed to function when they got there. As time went on, we humans got better at interplanetary flight. There was a learning curve. The adjacent figures show these curves (based on NASA data with NASA definitions of mission success). We learned very well. Our present ability to fix spacecraft in flight is best illustrated by the *Voyager* missions described earlier.

We see that it wasn't until about its thirty-fifth launch to the Moon or the planets that the cumulative U.S. mission success rate got as high as 50 percent. The Russians took about 50 launches to get there. Averaging the shaky start and the better recent performance, we find that both the United States and Russia have a cumulative *launch* success rate of about 80 percent. But the cumulative *mission* success rate is still under 70 percent for the U.S. and under 60 percent for the U.S.S.R./Russia. Equivalently, lunar and planetary missions have failed on average 30 or 40 percent of the time.

Missions to other worlds were from the beginning at the cutting edge of technology. They continue to be so today. They are designed with redundant subsystems, and operated by dedicated and experienced engineers, but they are not perfect. The amazing thing is not that we have done so poorly, but that we have done so well.

We don't know whether the *Mars Observer* failure was due to incompetence or just statistics. But we must expect a steady background of mission failures when we explore other worlds. No human lives are risked when a robot spacecraft is lost. Even if we were able to improve this success rate significantly, it would be far too costly. It is much better to take more risks and fly more spacecraft.

Knowing about irreducible risks, why do we these days fly only one spacecraft per mission? In 1962 *Mariner 1,* intended for Venus, fell into the Atlantic; the nearly identical *Mariner 2* became the human species' first successful planetary mission. *Mariner 3*

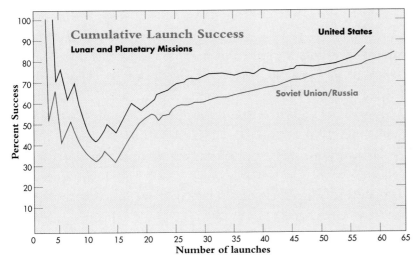

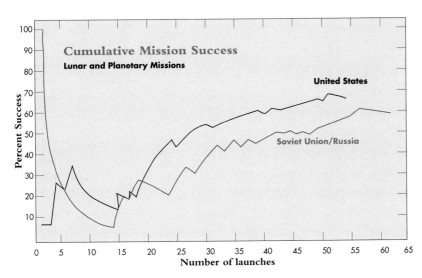

Success rate of U.S. and Soviet/Russian missions to the Moon and the planets—for successful launches, above, and for successful missions, below. There is clearly a steady learning curve, but mission failures are inevitable.

failed, and its twin *Mariner 4* became, in 1964, the first spacecraft to take close-up pictures of Mars. Or consider the 1971 *Mariner 8/Mariner 9* dual launch mission to Mars. *Mariner 8* was to map the planet. *Mariner 9* was to study the enigmatic seasonal and secular changes of surface markings. The spacecraft were otherwise identical. *Mariner 8* fell into the ocean. *Mariner 9* flew on to Mars and became the first spacecraft in human history to orbit another planet. It discovered the volcanos, the laminated terrain in the polar caps, the ancient river valleys, and the aeolian nature of the surface changes. It disproved the "canals." It mapped the planet pole to pole and revealed all the major geological features of Mars known to us today. It provided the first close-up observations of members of a whole class of small worlds (by targeting the Martian moons, Phobos and Deimos). If we had launched only *Mariner 8,* the endeavor would have been an unmitigated failure. With a dual launch it became a brilliant and historic success.

There were also two *Vikings,* two *Voyagers,* two *Vegas,* many pairs of *Veneras.* Why was only one *Mars Observer* flown? The standard answer is cost. Part of the reason it was so costly, though, is that it was planned to be launched by shuttle, which is an almost absurdly expensive booster for planetary missions—in this case too expensive for two *M. O.* launches. After many shuttle-connected delays and cost increases, NASA changed its mind and decided to launch *Mars Observer* on a *Titan* booster. This required an additional two-year delay and an adapter to mate the spacecraft to the new launch vehicle. If NASA had not been so intent on providing business for the increasingly uneconomic shuttle, we could have launched a couple of years earlier and maybe with two spacecraft instead of one.

But whether in single launches or in pairs, the spacefaring nations have clearly decided that the time is ripe to return robot explorers to Mars. Mission designs change; new nations enter the field; old nations find they no longer have the resources. Even already funded programs cannot always be relied upon. But current plans do reveal something of the intensity of effort and the depth of dedication.

As I write this book, there are tentative plans by the United States, Russia, France, Germany, Japan, Austria, Finland, Italy, Canada, the European Space Agency, and other entities for a coordinated robotic exploration of Mars. In the seven years between 1996 and 2003, a flotilla of some twenty-five spacecraft—most of them comparatively small and cheap—are to be sent from Earth to Mars. There will be no quick flybys among them; these are all long-duration orbiter and lander missions. The United States will refly all of the scientific instruments that were lost on *Mars Observer.* The Russian spacecraft will contain particularly ambitious experiments involving some twenty nations. Communications satellites will permit experimental stations anywhere on Mars to relay their data back to Earth. Penetrators screeching down from orbit will punch into the Martian soil, transmitting data from underground. Instrumented balloons and roving laboratories will wander over the sands of Mars. Some microrobots will weigh no more than a few pounds. Landing sites are being planned and coordinated. Instruments will be cross-calibrated. Data will be freely exchanged. There is every reason to think that in the coming years Mars and its mysteries will become increasingly familiar to the inhabitants of the planet Earth.

IN THE COMMAND CENTER on Earth, in a special room, you are helmeted and gloved. You turn your head to the left, and the cameras on the Mars robot rover turn to the left. You see, in very high def-

Highly maneuverable (see tracks) robot rover traverses the Martian surface. Its antenna is pointed at Earth. Painting by Michael Carroll.

inition and in color, what the cameras see. You take a step forward, and the rover walks forward. You reach out your arm to pick up something shiny in the soil, and the robot arm does likewise. The sands of Mars trickle through your fingers. The only difficulty with this remote reality technology is that all this must occur in tedious slow motion: The round-trip travel time of the up-link commands from Earth to Mars and the down-link data returned from Mars to Earth might take half an hour or more. But this is something we can learn to do. We can learn to contain our exploratory impatience if that's the price of exploring Mars. The rover can be made smart enough to deal with routine contingencies. Anything more challenging, and it makes a dead stop, puts itself into a safeguard mode, and radios for a very patient human controller to take over.

Conjure up roving, smart robots, each of them a small scientific laboratory, landing in the safe but dull places and wandering to view close-up some of that profusion of Martian wonders. Perhaps every day a robot would rove to its own horizon; each morning we would see close-up what had yesterday been only a distant eminence. The lengthening progress of a traverse route over the Martian landscape would appear on news programs and in schoolrooms. People would speculate on what will be found. Nightly newscasts from another planet, with their revelations of new terrains and new scientific findings, would make everyone on Earth a party to the adventure.

Then there's Martian virtual reality: The data sent back from Mars, stored in a modern computer, are fed into your helmet and gloves and boots. You are walking in an empty room on Earth, but to you you are on Mars: pink skies, fields of boulders, sand dunes stretching to the horizon where an immense volcano looms; you hear the sand crunching under your boots, you turn rocks over, dig a hole, sample the thin air, turn a corner, and come face to face with . . . whatever new discoveries we will make on Mars—all exact copies of what's on Mars, and all experienced from the safety of a virtual reality salon in your hometown. This is not *why* we explore Mars, but clearly we will need robot explorers to return the real reality before it can be reconfigured into virtual reality.

Especially with continuing investment in robotics and ma-

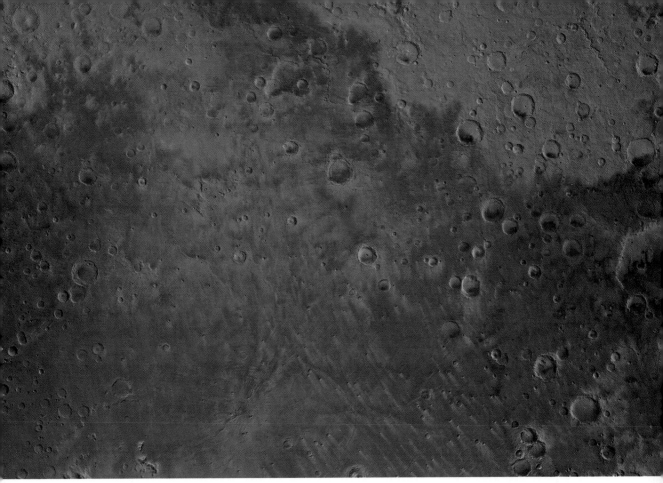

chine intelligence, sending humans to Mars can't be justified by science alone. And many more people can experience the virtual Mars than could possibly be sent to the real one. We can do very well with robots. If we're going to send people, we'll need a better reason than science and exploration.

In the 1980s, I thought I saw a coherent justification for human missions to Mars. I imagined the United States and the Soviet Union, the two Cold War rivals that had put our global civilization at risk, joining together in a far-seeing, high-technology endeavor that would give hope to people everywhere. I pictured a kind of *Apollo* program in reverse, in which cooperation, not competition, was the driving force, in which the two leading space-faring nations would together lay the groundwork for a major advance in human history—the eventual settlement of another planet.

The symbolism seemed so apt. The same technology that can propel apocalyptic weapons from continent to continent would

In this *Viking* photomosaic are several different terrain types. Especially at top right, dry outflow channels debouch into craters that must once have been filled with water. At bottom center is a field of crater-associated bright wind streaks. The nature of the dark material that fills many crater floors is unknown; it may simply be deposits of larger, less readily moved grains of the same composition as the bright deposits covering this province of Mars. Courtesy USGS/NASA.

enable the first human voyage to another planet. It was a choice of fitting mythic power: to embrace the planet named after, rather the madness ascribed to, the god of war.

We succeeded in interesting Soviet scientists and engineers in such a joint endeavor. Roald Sagdeev, then director of the Institute for Space Research of the Soviet Academy of Sciences in Moscow, was already deeply engaged in international cooperation on Soviet robotic missions to Venus, Mars, and Halley's Comet, long before the idea became fashionable. Projected joint use of the Soviet *Mir* space station and the *Saturn V*-class launch vehicle *Energiya* made cooperation attractive to the Soviet organizations that manufactured these items of hardware; they were otherwise having difficulty justifying their wares. Through a sequence of arguments (helping to bring the Cold War to an end being chief among them), then-Soviet leader Mikhail S. Gorbachev was convinced. During the December 1987 Washington summit, Mr. Gorbachev—asked what was the most important joint activity through which the two countries might symbolize the change in their relationship—unhesitatingly replied, "Let's go to Mars together."

But the Reagan Administration was not interested. Cooperating with the Soviets, acknowledging that certain Soviet technologies were more advanced than their American counterparts, making some American technology available to the Soviets, sharing credit, providing an alternative for the arms manufacturers—these were not to the Administration's liking. The offer was turned down. Mars would have to wait.

In only a few years, times have changed. The Cold War is over. The Soviet Union is no more. The benefit deriving from the two nations working together has lost some of its force. Other nations—especially Japan and the constituent members of the European Space Agency—have become interplanetary travelers. Many just and urgent demands are levied on the discretionary budgets of the nations.

But the *Energiya* heavy-lift booster still awaits a mission. The workhorse *Proton* rocket is available. The *Mir* space station—with a crew on board almost continuously—still orbits the Earth every hour and a half. Despite internal turmoil, the Russian space program continues vigorously. Cooperation between Russia and America in space is accelerating. A Russian cosmonaut, Sergei Krikalev, in 1994 flew on the shuttle *Discovery* (for the usual one-week shuttle mission duration; Krikalev had already logged 464 days aboard the *Mir* space station). U.S. astronauts will visit *Mir.* American instruments—including one to examine the oxidants thought to destroy organic molecules in the Martian soil—are to be carried by Russian space vehicles to Mars. *Mars Observer* was designed to serve as a relay station for landers in Russian Mars missions. The Russians have offered to include a U.S. orbiter in a forthcoming *Proton*-launched multipayload mission to Mars.

The American and Russian capabilities in space science and technology mesh; they interdigitate. Each is strong where the other is weak. This is a marriage made in heaven—but one that has been surprisingly difficult to consummate.

On September 2, 1993, an agreement to cooperate in depth was signed in Washington by Vice President Al Gore and Prime Minister Viktor Chernomyrdin. The Clinton Administration has ordered NASA to redesign the U.S. space station (called *Freedom* in the Reagan years) so it is in the same orbit as *Mir* and can be mated to it: Japanese and European modules will be attached, as will a Canadian robot arm. The designs have now evolved into what is called Space Station *Alpha,* involving almost all the space-faring nations. (China is the most notable exception.)

In return for U.S. space cooperation and an infusion of hard currency, Russia in effect agreed to halt its sale of ballistic missile components to other nations, and generally to exercise tight controls on its export of strategic weapons technology. In this way, space becomes once again, as it was at the height of the Cold War, an instrument of national strategic policy.

This new trend has, though, made some of the American aerospace industry and some key members of Congress profoundly uneasy. Without international competition, can we motivate such ambitious efforts? Does every Russian launch vehicle used cooperatively mean less support for the American aerospace industry? Can Americans rely on stable support and continuity of effort in joint projects with the Russians? (The Russians, of course, ask similar questions about the Americans.) But cooperative programs in the long term save money, draw upon the extraordinary scientific and engineering talent distributed over our planet, and provide inspiration about the global future. There may be fluctuations in national commitments. We are likely to take backward as well as forward steps. But the overall trend seems clear.

Despite growing pains, the space programs of the two former adversaries are beginning to join. It is now possible to foresee a world space station—not of any one nation but of the planet Earth—being assembled at 51° inclination to the equator and a few hundred miles up. A dramatic joint mission, called "Fire and Ice," is being discussed—in which a fast flyby would be sent to Pluto, the last unexplored planet; but to get there, a gravity assist from the Sun would be employed, in the course of which small probes would actually enter the Sun's atmosphere. And we seem to be on the threshold of a World Consortium for the scientific exploration of Mars. It very much looks as though such projects will be done cooperatively or not at all.

WHETHER THERE ARE VALID, cost–effective, broadly supportable reasons for people to venture to Mars is an open question. Certainly there is no consensus. The matter is treated in the next chapter.

I would argue that if we are not eventually going to send people to worlds as far away as Mars, we have lost the chief reason for a space station—a permanently (or intermittently) occupied human outpost in Earth orbit. A space station is far from an optimum platform for doing science—either looking down at the Earth, or looking out into space, or for utilizing microgravity (the very presence of astronauts messes things up). For military reconnaissance it is much inferior to robotic spacecraft. There are no compelling economic or manufacturing applications. It is expensive compared to robotic spacecraft. And of course it runs some risk of losing human lives. Every shuttle launch to help build or supply a space station has an estimated 1 or 2 percent chance of catastrophic failure. Previous civilian and military space activities have littered low Earth orbit with fast-moving debris—that sooner or later will collide with a space station (although, so far, *Mir* has had no failures from this hazard). A space station is also unnecessary for human exploration of the Moon. *Apollo* got there very well with no space station at all. With *Saturn V* or *Energiya* class launchers, it also may be possible to get to near-earth asteroids or even Mars without having to assemble the interplanetary vehicle on an orbiting space station.

A space station could serve inspirational and educational purposes, and it certainly can help to solidify relations among the spacefaring nations—particularly the United States and Russia. But the only substantive function of a space station, as far as I can see, is for long-duration spaceflight. How do humans behave in microgravity? How can we counter progressive changes in blood chemistry and an estimated 6 percent bone loss per year in zero gravity? (For a three- or four-year mission to Mars this adds up, if the travelers have to go at zero g.)

These are hardly questions in fundamental biology such as DNA or the evolutionary process; instead they address issues of applied human biology. It's important to know the answers, but only if we intend to go somewhere in space that's far away and takes a long time to get there. The only tangible and coherent goal of a space station is eventual human missions to near-Earth asteroids, Mars, and beyond. Historically NASA has been cautious about stating this fact clearly, probably for fear that members of Congress will throw up their hands in disgust, denounce the space station as the thin edge of an extremely expensive wedge, and declare the country unready to commit to launching people to Mars. In effect, then, NASA has kept quiet about what the space station is really for. And yet if we had such a space station, nothing would require us to go straight to Mars. We could use a space station to accumulate and refine the relevant knowledge, and take as long as we like to do so—so that when the time does come, when we are ready to go to the planets, we will have the background and experience to do so safely.

The *Mars Observer* failure, and the catastrophic loss of the space shuttle *Challenger* in 1986, remind us that there will be a certain irreducible chance of disaster in future human flights to Mars and elsewhere. The *Apollo 13* mission, which was unable to land

AT LEFT: oblique view toward the horizon over the Argyre Basin and the southern highlands.
AT RIGHT: a detached atmospheric dust layer can be seen, raised in the last major sand and dust storm. *Viking* orbiter images, courtesy JPL/NASA.

on the Moon and barely returned safely to Earth, underscores how lucky we've been. We cannot make perfectly safe autos or trains even though we've been at it for more than a century. Hundreds of thousands of years after we first domesticated fire, every city in the world has a service of firefighters biding their time until there's a blaze that needs putting out. In Columbus' four voyages to the New World, he lost ships left and right, including one third of the little fleet that set out in 1492.

If we are to send people, it must be for a very good reason— and with a realistic understanding that almost certainly we will lose lives. Astronauts and cosmonauts have always understood this. Nevertheless, there has been and will be no shortage of volunteers.

But why Mars? Why not return to the Moon? It's nearby, and we've proved we know how to send people there. I'm concerned that the Moon, close as it is, is a long detour, if not a dead end. We've been there. We've even brought some of it back. People have seen the Moon rocks, and, for reasons that I believe are fundamentally sound, they are bored by the Moon. It's a static, airless, waterless, black-sky, dead world. Its most interesting aspect perhaps is its cratered surface, a record of ancient catastrophic impacts, on the Earth as well as on the Moon.

Mars, by contrast, has weather, dust storms, its own moons, volcanos, polar ice caps, peculiar landforms, ancient river valleys, and evidence of massive climatic change on a once-Earthlike world. It holds some prospect of past or maybe even present life, and is the most congenial planet for future life—humans transplanted from Earth, living off the land. None of this is true for the Moon. Mars also has its own legible cratering history. If Mars, rather than the Moon, had been within easy reach, we would not have backed off from manned space flight.

Nor is the Moon an especially desirable test bed or way station for Mars. The Martian and lunar environments are very different, and the Moon is as distant from Mars as is the Earth. The machinery for Martian exploration can at least equally well be tested in Earth orbit, or on near-Earth asteroids, or on the Earth itself—in Antarctica, for instance.

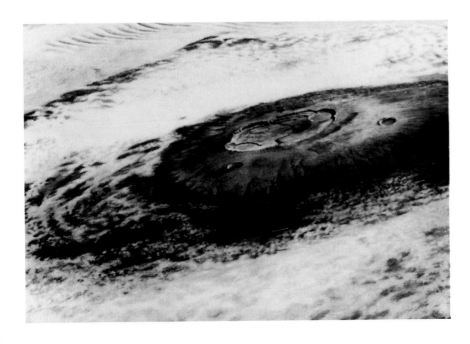

Olympus Mons sits high above surrounding clouds. Based on *Viking* imagery, courtesy JPL/NASA.

Japan has tended to be skeptical of the commitment of the United States and other nations to plan and execute major cooperative projects in space. This is at least one reason that Japan, more than any other spacefaring nation, has tended to go it alone. The Lunar and Planetary Society of Japan is an organization representing space enthusiasts in the government, universities, and major industries. As I write, the Society is proposing to construct and stock a lunar base entirely with robot labor. It is said to take about 30 years and to cost about a billion U.S. dollars a year (which would represent 7 percent of the present U.S. civilian space budget). Humans would arrive only when the base is fully ready. The use of robot construction crews under radio command from Earth is said to reduce the cost tenfold. The only trouble with the scheme, according to reports, is that other scientists in Japan keep asking, "What's it for?" That's a good question in every nation.

The first human mission to Mars is now probably too expensive for any one nation to pull off by itself. Nor is it fitting that such a historic step be taken by representatives of only a small fraction of the human species. But a cooperative venture among the United States, Russia, Japan, the European Space Agency—and perhaps other nations, such as China—might be feasible in the not too distant future. The international space station will have tested our ability to work together on great engineering projects in space.

The cost of sending a kilogram of something no farther away than low Earth orbit is today about the same as the cost of a kilogram of gold. This is surely a major reason we have yet to stride the ancient shorelines of Mars. Multistage chemical rockets are the means that first took us into space, and that's what we've been using ever since. We've tried to refine them, to make them safer, more reliable, simpler, cheaper. But that hasn't happened, or at least not nearly as quickly as many had hoped.

So maybe there's a better way: maybe single-stage rockets that can launch their payloads directly to orbit; maybe many small payloads shot from guns or rocket-launched from airplanes; maybe supersonic ramjets. Maybe there's something much better that we haven't thought of yet. If we can manufacture propellants for the return trip from the air and soil of our destination world, the difficulty of the voyage would be greatly eased.

Once we're up there in space, venturing to the planets, rocketry is not necessarily the best means to move large payloads around, even with gravity assists. Today, we make a few early rocket burns and later midcourse corrections, and coast the rest of the way. But there are promising ion and nuclear/electric propulsion systems by which a small and steady acceleration is exerted. Or, as the Russian space pioneer Konstantin Tsiolkovsky first envisioned, we could employ solar sails—vast but very thin films that catch sunlight and the solar wind, a caravel kilometers wide plying the void between the worlds. Especially for trips to Mars and beyond, such methods are far better than rockets.

As with most technologies, when something barely works, when it's the first of its kind, there's a natural tendency to improve it, develop it, exploit it. Soon there's such an institutional investment in the original technology, no matter how flawed, that it's very hard to move on to something better. NASA has almost no resources to pursue alternative propulsion technologies. That money would have to come out of near-term missions, missions which could provide concrete results and improve NASA's success record. Spending money on alternative technologies pays off a decade or two in the future. We tend to be very little interested in a decade or two in the future. This is one of the ways by which initial success can sow the seeds of ultimate failure; and is very similar to what sometimes happens in biological evolution. But sooner or later some nation—perhaps one without a huge investment in marginally effective technology—will develop effective alternatives.

Even before then, if we take a cooperative path, there will come a time—perhaps in the first decades of the new century and the new millennium—when an interplanetary spacecraft is assembled in Earth orbit, the progress in full view on the evening news. Astronauts and cosmonauts, hovering like gnats, guide and mate the prefabricated parts. Eventually the ship, tested and ready, is boarded by its international crew, and boosted to escape velocity. For the whole of the voyage to Mars and back, the lives of the

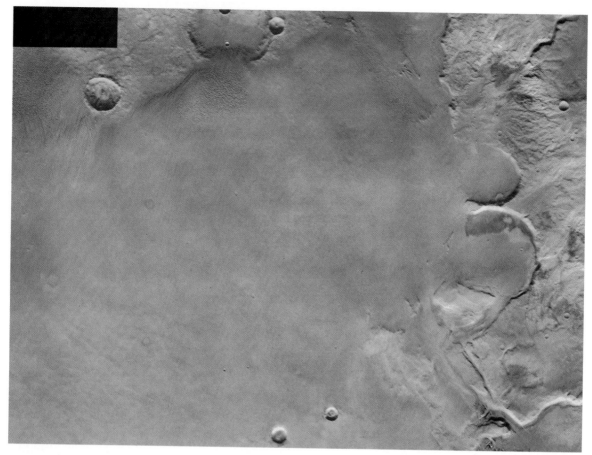

One of the many proposed landing sites for future rover missions to Mars: bland terrain near the ancient river valley Mangala Vallis. *Viking* photomosaic, courtesy USGS/NASA.

crew members depend on one another, a microcosm of our actual circumstances down here on Earth. Perhaps the first joint interplanetary mission with human crews will be only a flyby or orbit of Mars. Earlier, robot vehicles, with aerobraking, parachutes, and retrorockets, will have set gently down on the Martian surface to collect samples and return them to Earth, and to emplace supplies for future explorers. But whether or not we have compelling, coherent reasons, I am sure—unless we destroy ourselves first—that the day will come when we humans set foot on Mars. It is only a matter of when.

According to solemn treaty, signed in Washington and Moscow on January 27, 1967, no nation may lay claim to part or all of another planet. Nevertheless—for historical reasons that Columbus would have understood well—some people are concerned about who first sets foot on Mars. If this really worries us, we can arrange for the ankles of the crew members to be tied together as they alight in the gentle Martian gravity.

The crews would acquire new and previously sequestered samples, in part to search for life, in part to understand the past and future of Mars and Earth. They would experiment, for later expeditions, on extracting water, oxygen, and hydrogen from the rocks and the air and from the underground permafrost—to drink, to breathe, to power their machines and, as rocket fuel and oxidizer, to propel the return voyage. They would test Martian materials for eventual fabrication of bases and settlements on Mars.

And they would go exploring. When I imagine the early human exploration of Mars, it's always a roving vehicle, a little like a jeep, wandering down one of the valley networks, the crew with geological hammers, cameras, and analytic instruments at the ready. They're looking for rocks from ages past, signs of ancient cataclysms, clues to climate change, strange chemistries, fossils, or—most exciting and most unlikely—something alive. Their discoveries are televised back to Earth at the speed of light. Snuggled up in bed with the kids, you explore the ancient riverbeds of Mars.

CHAPTER 16

SCALING HEAVEN

Who, my friend, can scale heaven?
—*THE EPIC OF GILGAMESH*
(SUMER, THIRD MILLENNIUM B.C.)

What?, I sometimes ask myself in amazement: Our ancestors walked from East Africa to Novaya Zemlya and Ayers Rock and Patagonia, hunted elephants with stone spearpoints, traversed the polar seas in open boats 7,000 years ago, circumnavigated the Earth propelled by nothing but wind, walked the Moon a decade after entering space—and we're daunted by a voyage to Mars? But then I remind myself of the avoidable human suffering on Earth, how a few dollars can save the life of a child dying of dehydration, how many children we could save for the cost of a trip to Mars—and for the moment I change my mind. Is it unworthy to stay home, or unworthy to go? Or have I posed a false dichotomy? Isn't it possible to make a better life for everyone on Earth *and* to reach for the planets and the stars?

We had an expansive run in the '60s and '70s. You might have

OPPOSITE: Looking down on western Australia. The cargo bay of the space shuttle *Endeavor* is in foreground. Astronaut Story Musgrave approaches the snared Hubble Space Telescope (upright cylinder). In a flawless repair mission requiring five space "walks," the nearsighted telescope was fitted with corrective optics. Some of the breathtaking consequences can be seen elsewhere in this book. Courtesy Johnson Space Center/NASA.

thought, as I did then, that our species would be on Mars before the century was over. But instead, we've pulled inward. Robots aside, we've backed off from the planets and the stars. I keep asking myself: Is it a failure of nerve or a sign of maturity?

Maybe it's the most we could reasonably have expected. In a way it's amazing that it was possible at all: We sent a dozen humans on week-long excursions to the Moon. And we were given the resources to make a preliminary reconnaissance of the whole Solar System, out to Neptune anyway—missions that returned a wealth of data, but nothing of short-term, everyday, bread-on-the-table practical value. They lifted the human spirit, though. They enlightened us about our place in the Universe. It's easy to imagine skeins of historical causality in which there were no race to the Moon and no planetary program.

But it's also possible to imagine a much more serious devotion to exploration, because of which we would today have robot vehicles probing the atmospheres of all the Jovian planets and dozens of moons, comets, and asteroids; a network of automatic scientific stations emplaced on Mars would daily be reporting their findings; and samples from many worlds would be under examination in the laboratories of Earth—revealing their geology, chemistry, and perhaps even their biology. Human outposts might be already established on the near-Earth asteroids, the Moon, and Mars.

There were many possible historical paths. Our particular causality skein has brought us to a modest and rudimentary, although in many respects heroic, series of explorations. But it is far inferior to what might have been—and what may one day be.

"TO CARRY THE GREEN Promethean spark of Life with us into the sterile void and ignite there a firestorm of animate matter is the very destiny of our race," reads the brochure of something called the First Millennial Foundation. It promises, for $120 a year, "citizenship" in "space colonies—when the time comes." "Benefactors" who contribute more also receive "the undying gratitude of a starflung civilization, and their name carved on the monolith to be erected on the Moon." This represents one extreme in the continuum of enthusiasm for a human presence in space. The other extreme—better represented in Congress—questions why we should be in space at all, especially people rather than robots. The *Apollo* program was a "moondoggle," the social critic Amitai Etzioni once called it; with the Cold War over, there is no justification whatever, proponents of this orientation hold, for a manned space program. Where in this spectrum of policy options should we be?

Ever since the United States beat the Soviet Union to the Moon, a coherent, widely understood justification for humans in space seems to have vanished. Presidents and Congressional committees puzzle over what to do with the manned space program. What is it for? Why do we need it? But the exploits of the astronauts and the moon landings had elicited—and for good reason—the admiration of the world. It would be a rejection of that stunning American achievement, the political leaders tell themselves, to back off from manned spaceflight. Which President, which Congress wishes to be responsible for the end of the American space program? And in the former Soviet Union a similar argument is heard: Shall we abandon, they ask themselves, the one remaining high technology in which we are still world leaders? Shall we be faithless heirs of Konstantin Tsiolkovsky, Sergei Korolev, and Yuri Gagarin?

The first law of bureaucracy is to guarantee its own continuance. Left to its own devices, without clear instructions from above, NASA gradually devolved into a program that would maintain profits, jobs, and perquisites. Pork-barrel politics, with Congress playing a leading role, became an increasingly powerful force in the design and execution of missions and long-term goals. The bureaucracy ossified. NASA lost its way.

On July 20, 1989, the twentieth anniversary of the *Apollo 11* landing on the Moon, President George Bush announced a long-term direction for the U.S. space program. Called the Space Exploration Initiative (SEI), it proposed a sequence of goals including a U.S. space station, a return of humans to the Moon, and the first landing of humans on Mars. In a later statement, Mr. Bush set 2019 as the target date for the first footfall on that planet.

And yet the Space Exploration Initiative, despite clear direction from the top, foundered. Four years after it was mandated, it did not even have a NASA office dedicated to it. Small and inexpensive lunar robotic missions—that otherwise might well have been approved—were canceled by Congress because of guilt by association with SEI. What went wrong?

One problem was the timescale. SEI extended five or so presidential terms of office into the future (taking the average presidency as one and a half terms). That makes it easy for a president to attempt to commit his successors, but leaves in considerable doubt how reliable such a commitment might be. SEI contrasted dramatically with the *Apollo* program—which, it might have been conjectured at the time it began, could have triumphed when President Kennedy or his immediate political heir was still in office.

Second, there was concern about whether NASA, which had recently experienced great difficulty in safely lifting a few astronauts 200 miles above the Earth, could send astronauts on an arcing year-long trajectory to a destination 100 million miles away and bring them back alive.

Third, the program was conceived exclusively in nationalist terms. Cooperation with other nations was not fundamental to either design or execution. Vice President Dan Quayle, who had nominal responsibility for space, justified the space station as a demonstration that the United States was "the world's only superpower." But since the Soviet Union had an operational space station that was a decade ahead of the United States, Mr. Quayle's argument proved difficult to follow.

Finally, there was the question of where, in terms of practical politics, the money was supposed to come from. The costs of getting the first humans to Mars had been variously estimated, ranging as high as $500 billion.

Of course, it's impossible to predict costs before you have a mission design. And the mission design depends on such matters as the size of the crew; the extent to which you take mitigating steps against solar and cosmic radiation hazards, or zero gravity; and what other risks you are willing to accept with the lives of the men and women on board. If every crew member has one essential specialty, what happens if one of them falls ill? The larger the crew, the more reliable the backups. You would almost certainly not send a full-time oral surgeon, but what happens if you need root canal work and you're a hundred million miles from the nearest dentist? Or could it be done by an endodontist on Earth, using telepresence?

Wernher von Braun was the Nazi-American engineer who, more than anyone else, actually took us into space. His 1952 book *Das Marsprojekt* envisioned a first mission with 10 interplanetary spacecraft, 70 crew members, and 3 "landing boats." Redundancy was uppermost in his mind. The logistical requirements, he wrote, "are no greater than those for a minor military operation extending over a limited theater of war." He meant to "explode once and for all the theory of the solitary space rocket and its little band of bold interplanetary adventurers," and appealed to Columbus' three ships without which "history tends to prove that he might never have returned to Spanish shores." Modern Mars mission designs have ignored this advice. They are much less ambitious than von Braun's, typically calling for one or two spacecraft crewed by three to eight astronauts, with another robotic cargo ship or two. The solitary rocket and the little band of adventurers are still with us.

Other uncertainties affecting mission design and cost include whether you pre-emplace supplies from Earth and launch humans to Mars only after the supplies are safely landed; whether you can use Martian materials to generate oxygen to breathe, water to drink, and rocket propellants to get home; whether you land using the thin Martian atmosphere for aerobraking; the degree of redundancy in equipment thought prudent; the extent to which you use closed ecological systems or just depend on the food, water, and waste disposal facilities you've brought from Earth; the design of roving vehicles for the crew to explore the Martian landscape; and how much equipment you're willing to carry to test our ability to live off the land in later voyages.

Until such questions are decided, it's absurd to accept any figure for the cost of the program. On the other hand, it was equally clear that SEI would be extremely expensive. For all these reasons, the program was a nonstarter. It was stillborn. There was no effective attempt by the Bush Administration to spend political capital to get SEI going.

The lesson to me seems clear: There may be no way to send humans to Mars in the comparatively near future—despite the fact that it is entirely within our technological capability. Governments do not spend these vast sums just for science, or merely to explore. They need another purpose, and it must make real political sense.

It may be impossible to go just yet, but when it is possible, the mission, I think, must be international from the start, with costs and responsibilities equitably shared and the expertise of many nations tapped; the price must be reasonable; the time from approval to launch must fit within practical political timescales; and the space agencies concerned must demonstrate their ability to muster pioneering exploratory missions with human crews safely, on time, and on budget. If it were possible to imagine such a mission for less than $100 billion, and for a time from approval to launch less than 15 years, maybe it would be feasible. (In terms of cost, this would represent only a fraction of the annual civilian space budgets of the present spacefaring nations.) With aerobraking and manufacturing fuel and oxygen for the return trip out of Martian air, it's now beginning to look as if such a budget and such a timescale might actually be realistic.

The cheaper and quicker the mission is, necessarily the more risk we must be willing to take with the lives of the astronauts and cosmonauts aboard. But as is illustrated, among countless examples, by the samurai of medieval Japan, there are always competent volunteers for highly dangerous missions in what is perceived as a great cause. No budget, no timeline can be really reliable when we attempt to do something on so grand a scale, something that has never been done before. The more leeway we ask, the greater is the cost and the longer it takes to get there. Finding the right compromise between political feasibility and mission success may be tricky.

IT'S NOT ENOUGH to go to Mars because some of us have dreamt of doing so since childhood, or because it seems to us the obvious long-term exploratory goal for the human species. If we're talking about spending this much money, we must justify the expense.

There are now other matters—clear, crying national needs—that cannot be addressed without major expenditures; at the same time, the discretionary federal budget has become painfully constrained. Disposal of chemical and radioactive poisons, energy efficiency, alternatives to fossil fuels, declining rates of technological innovation, the collapsing urban infrastructure, the AIDS epidemic, a witches' brew of cancers, homelessness, malnutrition, infant mortality, education, jobs, health care—there is a painfully long list. Ignoring them will endanger the well-being of the nation. A similar dilemma faces all the spacefaring nations.

Nearly every one of these matters could cost hundreds of billions of dollars or more to address. Fixing infrastructure will cost several trillion dollars. Alternatives to the fossil-fuel economy clearly represent a multitrillion-dollar investment worldwide, if we can do it. These projects, we are sometimes told, are beyond our ability to pay. How then can we afford to go to Mars?

If there were 20 percent more discretionary funds in the U.S. federal budget (or the budgets of the other spacefaring nations), I probably would not feel so conflicted about advocating sending humans to Mars. If there were 20 percent less, I don't think the most diehard space enthusiast would be urging such a mission. Surely there is some point at which the national economy is in such dire straits that sending people to Mars is unconscionable. The question is where we draw the line. Plainly such a line exists, and every participant in these debates should stipulate where that line should be drawn, what fraction of the gross national product for space is too much. I'd like the same thing done for "defense."

Public opinion polls show that many Americans think the NASA budget is about equal to the defense budget. In fact, the entire NASA budget, including human and robotic missions and aeronautics, is about 5 percent of the U.S. defense budget. How much spending for defense actually weakens the country? And even if NASA were cancelled altogether, would we free up what is needed to solve our national problems?

HUMAN SPACEFLIGHT in general—to say nothing of expeditions to Mars—would be much more readily supportable if, as in the fifteenth-century arguments of Columbus and Henry the Navigator, there were a profit lure.★ Some arguments have been advanced. The high vacuum or low gravity or intense radiation environment of near-Earth space might be utilized, it is said, for commercial benefit. All such proposals must be challenged by this question: Could comparable or better products be manufactured down here on Earth if the development money made available were comparable to what is being poured into the space program? Judging by how little money corporations have been willing to invest in such technology—apart from the entities building the rockets and spacecraft themselves—the prospects, at least at present, seem to be not very high.

The notion that rare materials might be available elsewhere is tempered by the fact that freightage is high. There may, for all we know, be oceans of petroleum on Titan, but transporting it to Earth will be expensive. Platinum-group metals may be abundant in certain asteroids. If we could move these asteroids into orbit around the Earth, perhaps we could conveniently mine them. But at least for the foreseeable future this seems dangerously imprudent, as I describe later in this book.

In his classic science fiction novel *The Man Who Sold the Moon,* Robert Heinlein imagined the profit motive as the key to space travel. He hadn't foreseen that the Cold War would sell the Moon. But he did recognize that an honest profit argument would be difficult to come by. Heinlein envisioned, therefore, a scam in which the lunar surface was salted with diamonds so later explorers could breathlessly discover them and initiate a diamond rush. We've since returned samples from the Moon, though, and there is not a hint of commercially interesting diamonds there.

★ Even then it wasn't easy. The Portuguese chronicler Gomes Eanes de Zurara reported this assessment by Prince Henry the Navigator: "It seemed to the Lord Infante that if he or some other lord did not endeavor to gain that knowledge, no mariners nor merchants would ever dare to attempt it, for it is clear that none of them ever trouble themselves to sail to a place where there is not a sure and certain hope of profit."

However, Kiyoshi Kuramoto and Takafumi Matsui of the University of Tokyo have studied how the central iron cores of Earth, Venus, and Mars formed, and find that the Martian mantle (between crust and core) should be rich in carbon—richer than that of the Moon or Venus or Earth. Deeper than 300 kilometers, the pressures should transform carbon into diamond. We know that Mars has been geologically active over its history. Material from great depth will occasionally be extruded up to the surface, and not just in the great volcanos. So there does seem to be a case for diamonds on other worlds—on Mars, and not the Moon. In what quantities, of what quality and size, and in which locales we do not yet know.

The return to Earth of a spacecraft stuffed with gorgeous multicarat diamonds would doubtless depress prices (as well as the shareholders of the de Beers and General Electric corporations). But because of the ornamental and industrial applications of diamonds, perhaps there is a lower limit below which prices will not go. Conceivably, the affected industries might find cause to promote the early exploration of Mars.

The idea that Martian diamonds will pay for exploring Mars is at best a very long shot, but it's an example of how rare and valuable substances may be discoverable on other worlds. It would be foolish, though, to count on such contingencies. If we seek to justify missions to other worlds, we'll have to find other reasons.

BEYOND DISCUSSIONS OF PROFITS and costs, even reduced costs, we must also describe benefits, if they exist. Advocates of human missions to Mars must address whether, in the long term, missions up there are likely to mitigate any of the problems down here. Consider now the standard set of justifications and see if you find them valid, invalid, or indeterminate:

Human missions to Mars would spectacularly improve our knowledge of the planet, including the search for present and past life. The program is likely to clarify our understanding of the environment of our own planet, as robotic missions have already begun to do. The history of our civilization shows that the pursuit of basic knowledge is the way the most significant practical advances come about. Opinion polls suggest that the most popular reason for "exploring space" is "increased knowledge." But are humans in space essential to achieve this goal? Robotic missions, given high national priority and equipped with improved machine intelligence, seem to me entirely capable of answering, as well as astronauts can, all the questions we need to ask—and at maybe 10 percent the cost.

It is alleged that "spinoff" will transpire—huge technological benefits that would otherwise fail to come about—thereby improving our international competitiveness and the domestic economy. But this is an old argument: Spend $80 billion (in contemporary money) to send *Apollo* astronauts to the Moon, and we'll throw in a free stickless frying pan. Plainly, if we're after frying pans, we can invest the money directly and save almost all of that $80 billion.

The argument is specious for other reasons as well, one of which is that DuPont's Teflon technology long antedated *Apollo*. The same is true of cardiac pacemakers, ballpoint pens, Velcro, and other purported spinoffs of the *Apollo* program. (I once had the opportunity to talk with the inventor of the cardiac pacemaker, who himself nearly had a coronary accident describing the injustice of what he perceived as NASA taking credit for his device.) If there are technologies we urgently need, then spend the money and develop them. Why go to Mars to do it?

Of course it would be impossible for so much new technology as NASA requires to be developed and not have some spillover into the general economy, some inventions useful down here. For example, the powdered orange juice substitute Tang was a product of the manned space program, and spinoffs have occurred in cordless tools, implanted cardiac defibrillators, liquid-cooled garments, and digital imaging—to name a few. But they hardly justify human voyages to Mars or the existence of NASA.

We could see the old spinoff engine wheezing and puffing in the waning days of the Reagan-era Star Wars office. Hydrogen bomb–driven X-ray lasers on orbiting battle stations will help perfect laser surgery, they told us. But if we need laser surgery, if it's a high national priority, by all means let's allocate the funds to develop it. Just leave Star Wars out of it. Spinoff justifications constitute an admission that the program can't stand on its own two feet, cannot be justified by the purpose for which it was originally sold.

Once upon a time it was thought, on the basis of econometric models, that for every dollar invested in NASA many dollars were pumped into the U.S. economy. If this multiplier effect applied more to NASA than to most government agencies, it would provide a potent fiscal and social justification for the space program. NASA supporters were not shy about appealing to this argument. But a 1994 Congressional Budget Office study found it to be a delusion. While NASA spending benefits some production segments of the U.S. economy—especially the aerospace industry—there is no preferential multiplier effect. Likewise, while NASA spending certainly creates or maintains jobs and profits, it does so no more efficiently than many other government agencies.

Then there's education, an argument that has proved from time to time very attractive in the White House. Doctorates in science peaked somewhere around the time of *Apollo 11,* maybe even with the proper phase lag after the start of the *Apollo* program. The cause-and-effect relationship is perhaps undemonstrated, although not implausible. But so what? If we're interested in improving education, is going to Mars the best route? Think of what we could do with $100 billion for teacher training and salaries, school laboratories and libraries, scholarships for disadvantaged students, research facilities, and graduate fellowships. Is it really true that the best way to promote science education is to go to Mars?

Another argument is that human missions to Mars will occupy the military-industrial complex, diffusing the temptation to use its considerable political muscle to exaggerate external threats and pump up defense funding. The other side of this particular coin is that by going to Mars we maintain a standby technological capacity that might be important for future military contingencies. Of course, we might simply ask those guys to do something directly useful for the civilian economy. But as we saw in the 1970s with Grumman buses and Boeing/Vertol commuter trains, the aerospace industry experiences real difficulty in producing competitively for the civilian economy. Certainly a tank may travel 1,000 miles a year and a bus 1,000 miles a week, so the basic designs must be different. But on matters of reliability at least, the Defense Department seems to be much less demanding.

Cooperation in space, as I've already mentioned, is becoming an instrument of international cooperation—for example, in slowing the proliferation of strategic weapons to new nations. Rockets decommissioned because of the end of the Cold War might be gainfully employed in missions to Earth orbit, the Moon, the planets, asteroids, and comets. But all this can be accomplished without human missions to Mars.

Other justifications are offered. It is argued that the ultimate solution to world energy problems is to strip-mine the Moon, return the solar-wind-implanted helium-3 back to Earth, and use it in fusion reactors. What fusion reactors? Even if this were possible, even if it were cost-effective, it is a technology 50 or 100 years away. Our energy problems need to be solved at a less leisurely pace.

Even stranger is the argument that we have to send human beings into space in order to solve the world population crisis. But some 250,000 more people are born than die every day—which means that we would have to launch 250,000 people per day into space to maintain world population at its present levels. This appears to be beyond our present capability.

I RUN THROUGH such a list and try to add up the pros and cons, bearing in mind the other urgent claims on the federal budget. To me, the argument so far comes down to this question: Can the sum of a large number of individually inadequate justifications add up to an adequate justification?

I don't think any of the items on my list of purported justifications is demonstrably worth $500 billion or even $100 billion, certainly not in the short term. On the other hand, most of them are worth something, and if I have five items each worth $20 billion, maybe it adds up to $100 billion. If we can be clever about reducing costs and making true international partnerships, the justifications become more compelling.

Until a national debate on this topic has transpired, until we have a better idea of the rationale and the cost/benefit ratio of human missions to Mars, what should we do? My suggestion is that we pursue research and development projects that can be justified on their own merits or by their relevance to other goals, but that can also contribute to human missions to Mars should we later decide to go. Such an agenda would include:

- U.S. astronauts on the Russian space station *Mir* for joint flights of gradually increasing duration, aiming at one to two years, the Mars flight time.
- Configuration of the international space station so its principal function is to study the long-term effects of the space environment on humans.
- Early implementation of a rotating or tethered "artificial gravity" module on the international space station, for other animals and then for humans.
- Enhanced studies of the Sun, including a distributed set of robot probes in orbit about the Sun, to monitor solar activity and give the earliest possible warning to astronauts of hazardous "solar flares"—mass ejections of electrons and protons from the Sun's corona.
- U.S./Russian and multilateral development of *Energiya* and *Proton* rocket technology for the U.S. and international space programs. Although the United States is unlikely to depend primarily on a Soviet booster, *Energiya* has roughly the lift of the *Saturn V* that sent the *Apollo* astronauts to the Moon. The United States let the *Saturn V* assembly line die, and it cannot readily be resuscitated. *Proton* is the most reliable large booster now in service. Russia is eager to sell this technology for hard currency.
- Joint projects with NASDA (the Japanese space agency) and Tokyo University, the European Space Agency, and the Russian Space Agency, along with Canada and other nations. In most cases these should be equal partnerships, not the United

Shuttle and *Mir* rendezvous and dock in the first stage of U.S./Russian cooperation—with other international partners—in the creation of an international space station. Art by John Frassanito and Associates, courtesy NASA.

States insisting on calling the shots. For the robotic exploration of Mars, such programs are already under way. For human flight, the chief such activity is clearly the international space station. Eventually, we might muster joint simulated planetary missions in low Earth orbit. One of the principal objectives of these programs should be to build a tradition of cooperative technical excellence.

• Technological development—using state-of-the-art robotics and artificial intelligence—of rovers, balloons, and aircraft for

the exploration of Mars, and implementation of the first international return sample mission. Robotic spacecraft that can
return samples from Mars can be tested on near-Earth asteroids and the Moon. Samples returned from carefully selected
regions of the Moon can have their ages determined and contribute in a fundamental way to our understanding of the
early history of the Earth.

• Further development of technologies to manufacture fuel and
oxidizer out of Martian materials. In one estimate, based on a
prototype instrument designed by Robert Zubrin and colleagues at the Martin Marietta Corporation, several kilograms
of Martian soil can be automatically returned to Earth using a

Two future spacecraft rendezvous
above the lunar surface. NASA
artwork by Pat Rawlings/SAIC.

modest and reliable Delta launch vehicle, all for no more than a song (comparatively speaking).

- Simulations on Earth of long-duration trips to Mars, concentrating on potential social and psychological problems.
- Vigorous pursuit of new technologies such as constant-thrust propulsion to get us to Mars quickly; this may be essential if the radiation or microgravity hazards make one-year (or longer) flight times too risky.
- Intensive study of near-Earth asteroids, which may provide superior intermediate-timescale objectives for human exploration than does the Moon.
- A greater emphasis on science—including the fundamental sciences behind space exploration, and the thorough analysis of data already obtained—by NASA and other space agencies.

These recommendations add up to a fraction of the full cost of a human mission to Mars and—spread out over a decade or so and done jointly with other nations—a fraction of current space budgets. But, if implemented, they would help us to make accurate cost estimates and better assessment of the dangers and benefits. They would permit us to maintain vigorous progress toward human expeditions to Mars without premature commitment to any specific mission hardware. Most, perhaps all, of these recommendations have other justifications, even if we were sure we'd be unable to send humans to any other world in the next few decades. And a steady drumbeat of accomplishments increasing the feasibility of human voyages to Mars would—in the minds of many at least—combat widespread pessimism about the future.

THERE'S SOMETHING MORE. There's a set of less tangible arguments, many of which, I freely admit, I find attractive and resonant. Spaceflight speaks to something deep inside us—many of us, if not all. An emerging cosmic perspective, an improved understanding of our place in the Universe, a highly visible program affecting our view of ourselves might clarify the fragility of our planetary environment and the common peril and responsibility of all the nations

A Mars robot vehicle (foreground) being tested on the surface of the Moon. NASA artwork by Pat Rawlings/SAIC.

and peoples of Earth. And human missions to Mars would provide hopeful prospects, rich in adventure, for the wanderers among us, especially the young. Even vicarious exploration has social utility.

I repeatedly find that when I give talks on the future of the space program—to universities, business and military groups, professional organizations—the audiences are much less patient with practical, real-world political and economic obstacles than I. They long to sweep away the impediments, to recapture the glory days of *Vostok* and *Apollo,* to get on with it and once more tread other worlds. We did it before; we can do it again, they say. But, I caution myself, those who attend such talks are self-selected space enthusiasts.

In 1969, less than half the American people thought the

A future electric propulsion vehicle approaches Mars. NASA artwork by Pat Rawlings/SAIC.

Apollo program was worth the cost. But on the twenty-fifth anniversary of the Moon landing, the number had risen to two thirds. Despite its problems, NASA was rated as doing a good-to-excellent job by 63 percent of Americans. With no reference to cost, 55 percent of Americans (according to a CBS News poll) favored "the United States sending astronauts to explore Mars." For young adults, the figure was 68 percent. I think "explore" is the operative word.

It is no accident that, whatever their human flaws, and however moribund the human space program has become (a trend that the Hubble Space Telescope repair mission may have helped to reverse), astronauts and cosmonauts are still widely regarded as heroes of our species. A scientific colleague tells me about a recent trip to the New Guinea highlands where she visited a stone age culture hardly contacted by Western civilization. They were ignorant of wristwatches, soft drinks, and frozen food. But they knew about *Apollo 11*. They knew that humans had walked on the Moon. They knew the names of Armstrong and Aldrin and Collins. They wanted to know who was visiting the Moon these days.

Projects that are future-oriented, that, despite their political difficulties, can be completed only in some distant decade are continuing reminders that there *will* be a future. Winning a foothold on other worlds whispers in our ears that we're more than Picts or Serbs or Tongans: We're humans.

Exploratory spaceflight puts scientific ideas, scientific thinking, and scientific vocabulary in the public eye. It elevates the general level of intellectual inquiry. The idea that we've now understood something never grasped by anyone who ever lived before—that exhilaration, especially intense for the scientists involved, but perceptible to nearly everyone—propagates through the society, bounces off walls, and comes back at us. It encourages us to address problems in other fields that have also never before been solved. It increases the general sense of optimism in the society. It gives currency to critical thinking of the sort urgently needed if we are to solve hitherto intractable social issues. It helps stimulate a new generation of scientists. The more science in the media—especially if methods are described, as well as conclusions and implications—the healthier, I believe, the society is. People everywhere hunger to understand.

WHEN I WAS A CHILD, my most exultant dreams were about flying—not in some machine, but all by myself. I would be skipping or hopping, and slowly I could pull my trajectory higher. It would take longer to fall back to the ground. Soon I would be on such a high arc that I wouldn't come down at all. I would alight like a gargoyle in a niche near the pinnacle of a skyscraper, or gently settle down on a cloud. In the dream—which I must have had in its many variations at least a hundred times—achieving flight required a certain cast of mind. It's impossible to describe it in words, but I can remember what it was like to this day. You did something inside your head and at the pit of your stomach, and then you could lift yourself up by an effort of will alone, your limbs hanging limply. Off you'd soar.

I know many people have had similar dreams. Maybe most people. Maybe everyone. Perhaps it goes back 10 million years or more, when our ancestors were gracefully flinging themselves from branch to branch in the primeval forest. A wish to soar like the birds motivated many of the pioneers of flight, including Leonardo da Vinci and the Wright brothers. Maybe that's part of the appeal of spaceflight, too.

In orbit about any world, or in interplanetary flight, you are literally weightless. You can propel yourself to the spacecraft ceiling with a slight push off the floor. You can go tumbling through the air down the long axis of the spacecraft. Humans experience weightlessness as joy; this has been reported by almost every astronaut and cosmonaut. But because spacecraft are still so small, and because space "walks" have been done with extreme caution, no human has yet enjoyed this wonder and glory: propelling yourself by an almost imperceptible push, with no machinery driving you, untethered, high up into the sky, into the blackness of interplanetary space. You become a living satellite of the Earth, or a human planet of the Sun.

Planetary exploration satisfies our inclination for great enterprises and wanderings and quests that has been with us since our days as hunters and gatherers on the East African savannahs a million years ago. By chance—it is possible, I say, to imagine many skeins of historical causality in which this would not have transpired—in our age we are able to begin again.

OPPOSITE: Two astronauts or cosmonauts approach the *Viking 2* lander on Utopia Planitia, Mars. In the early morning light, it is hard to see which flag is being carried. Perhaps it is the flag of Earth. NASA artwork by Pat Rawlings/SAIC.

LEFT: The most exciting, but perhaps least likely, outcome of Mars exploration: the discovery of an unknown ancient civilization. Here, a globe of Earth as it appeared 250 million years ago is disinterred, inscribed in an unknown hieroglyphic writing. Painting by Pat Rawlings; copyright Pat Rawlings 1991.

Exploring other worlds employs precisely the same qualities of daring, planning, cooperative enterprise, and valor that mark the finest in the military tradition. Never mind the night launch of an *Apollo* spacecraft bound for another world. That makes the conclusion foregone. Witness mere F–14s taking off from adjacent flight decks, gracefully canting left and right, afterburners flaming, and there's something that sweeps you away—or at least it does me. And no amount of knowledge of the potential abuses of carrier task forces can affect the depth of that feeling. It simply speaks to another part of me. It doesn't want recriminations or politics. It just wants to fly.

"I . . . had ambition not only to go farther than anyone had done before," wrote Captain James Cook, the eighteenth-century explorer of the Pacific, "but as far as it was possible for man to go." Two centuries later, Yuri Romanenko, on returning to Earth after what was then the longest space flight in history, said "The Cosmos is a magnet . . . Once you've been there, all you can think of is how to get back."

Even Jean-Jacques Rousseau, no enthusiast of technology, felt it:

> The stars are far above us; we need preliminary instruction, instruments and machines, which are like so many immense ladders enabling us to approach them and bring them within our grasp.

"The future possibilities of space-travel," wrote the philosopher Bertrand Russell in 1959,

> which are now left mainly to unfounded fantasy, could be more soberly treated without ceasing to be interesting and could show to even the most adventurous of the young that a world without war need not be a world without adventurous and hazardous glory.[*] To this kind of contest there is no limit. Each victory is only a prelude to another, and no boundaries can be set to rational hope.

---

[*] Russell's phrase is noteworthy: "adventurous and hazardous glory." Even if we could make human spaceflight risk-free—and of course we cannot—it might be counterproductive. The hazard is an inseparable component of the glory.

In the long run, these—more than any of the "practical" justifications considered earlier—may be the reasons we will go to Mars and other worlds. In the meantime, the most important step we can take toward Mars is to make significant progress on Earth. Even modest improvements in the social, economic, and political problems that our global civilization now faces could release enormous resources, both material and human, for other goals.

There's plenty of housework to be done here on Earth, and our commitment to it must be steadfast. But we're the kind of species that needs a frontier—for fundamental biological reasons. Every time humanity stretches itself and turns a new corner, it receives a jolt of productive vitality that can carry it for centuries.

There's a new world next door. And we know how to get there.

covered was an extremely thin ring that surrounds Saturn at its equator but touches it nowhere. In some years, because of the changing orbital positions of Earth and Saturn, the ring had been seen edge-on and, because of its thinness, it seemed to disappear. In other years, it had been viewed more face-on, and the "ears" grew bigger. But what does it mean that there's a ring around Saturn? A thin, flat, solid plate with a hole cut out for the planet to fit into? Where does *that* come from?

This line of inquiry will shortly take us to world-shattering collisions, to two quite different perils for our species, and to a reason——beyond those already described—that we must, for our very survival, be out there among the planets.

We now know that the rings (emphatically plural) of Saturn are a vast horde of tiny ice worlds, each on its separate orbit, each bound to Saturn by the giant planet's gravity. In size, these worldlets range from particles of fine dust to houses. None is big enough to photograph even from close flybys. Spaced out in an exquisite set of fine concentric circles, something like the grooves on a phonograph record (which in reality make, of course, a spiral), the rings were first revealed in their true majesty by the two *Voyager* spacecraft in their 1980/81 flybys. In our century, the Art Deco rings of Saturn have become an icon of the future.

At a scientific meeting in the late 1960s, I was asked to summarize the outstanding problems in planetary science. One, I suggested, was the question of why, of all the planets, only Saturn had rings. This, *Voyager* discovered, is a nonquestion. *All four* giant planets in our Solar System—Jupiter, Saturn, Uranus, and Neptune—in fact have rings. But no one knew it then.

Each ring system has distinctive features. Jupiter's is tenuous and made mainly of dark, very small particles. The bright rings of Saturn are composed mainly of frozen water; there are thousands of separate rings here, some twisted, with strange, dusky, spokelike markings forming and dissipating. The dark rings of Uranus seem to be composed of elemental carbon and organic molecules—something like charcoal or chimney soot; Uranus has nine main rings, a few of which sometime seem to "breathe," expanding and contracting. Neptune's rings are the most tenuous of all, varying so

INTERPL

It is a law of nature that F
their proper places a

There was something funny about Saturn. When, in
Galileo used the world's first astronomical telesco
view the planet—then the most distant world know
found two appendages, one on either side. He likened th
"handles." Other astronomers called them "ears." The C
holds many wonders, but a planet with jug ears is disn
Galileo went to his grave with this bizarre matter unresolve

As the years passed, observers found the ears . . . well,
and waning. Eventually, it became clear that what Galileo h

The rings of Saturn are thinner compared to their width than is the piece of paper on which these words are printed. Were you to see the rings exactly edge-on, they would almost disappear. *Voyager* color-enhanced image, courtesy JPL/NASA.

Intricate detail among the hundreds of rings circling Saturn. Written in these details is a history of past catastrophes. *Voyager* image, courtesy JPL/NASA.

much in thickness that, when detected from Earth, they appear only as arcs and incomplete circles. A number of rings seem to be maintained by the gravitational tugs of two shepherd moons, one a little nearer and the other a little farther from the planet than the ring. Each ring system displays its own, appropriately unearthly, beauty.

How do rings form? One possibility is tides: If an errant world passes close to a planet, the interloper's near side is gravitationally pulled toward the planet more than its far side; if it comes close enough, if its internal cohesion is low enough, it can be literally torn to pieces. Occasionally we see this happening to comets as they pass too close to Jupiter, or the Sun. Another possibility, emerging from the *Voyager* reconnaissance of the outer Solar System, is this: Rings are made when worlds collide and moons are smashed to smithereens. Both mechanisms may have played a role.

The space between the planets is traversed by an odd collection of rogue worldlets, each in orbit about the Sun. A few are as big as a county or even a state; many more have surface areas like those of a village or a town. More little ones are found than big ones, and they range in size down to particles of dust. Some of them travel on long, stretched-out elliptical paths, which make them periodically cross the orbit of one or more planets.

Occasionally, unluckily, there's a world in the way. The collision can shatter and pulverize both the interloper and the moon that's hit (or at least the region around ground zero). The resulting debris—ejected from the moon but not so fast-moving as to escape from the planet's gravity—may form, for a time, a new ring. It's made of whatever the colliding bodies were made of, but usually more of the target moon than the rogue impactor. If the colliding worlds are icy, the net result will be rings of ice particles; if they're made of organic molecules, the result will be rings of organic particles (which will slowly be processed by radiation into carbon). All the mass in the rings of Saturn is no more than would result from the complete impact pulverization of a single icy moon. The disintegration of small moons can likewise account for the ring systems of the three other giant planets.

Unless it's very close to its planet, a shattered moon gradually reaccumulates (or at least a fair fraction of it does). The pieces, big

View of the surface of an asteroid approaching the north pole of Mars. Painting by William K. Hartmann.

and small, still in approximately the same orbit as the moon was before the impact, fall together helter-skelter. What used to be a piece of the core is now at the surface, and vice versa. The resulting hodgepodge surfaces might seem very odd. Miranda, one of the moons of Uranus, looks disconcertingly jumbled and may have had such an origin.

The American planetary geologist Eugene Shoemaker proposes that many moons in the outer Solar System have been annihilated and re-formed—not just once but several times each over the 4.5 billion years since the Sun and the planets condensed out of interstellar gas and dust. The picture emerging from the *Voyager* reconnaissance of the outer Solar System is of worlds whose placid and lonely vigils are spasmodically troubled by interlopers from space; of world-shattering collisions; and of moons re-forming

Airbrushed rendition of the Martian moon Phobos, thought to be a captured Main Belt asteroid. It is about 10 kilometers long—the size of the world that 65 million years ago ended the age of the dinosaurs on Earth. Courtesy USGS.

from debris, reconstituting themselves like phoenixes from their own ashes.

But a moon that lives very close to a planet cannot re-form if it is pulverized—the gravitational tides of the nearby planet prevent it. The resulting debris, once spread out into a ring system, might be very long-lived—at least by the standard of a human lifetime. Perhaps many of the small, inconspicuous moons now orbiting the giant planets will one day blossom forth into vast and lovely rings.

These ideas are supported by the appearance of a number of satellites in the Solar System. Phobos, the inner moon of Mars, has a large crater named Stickney; Mimas, an inner moon of Saturn, has a big one named Herschel. These craters—like those on our own Moon and, indeed, throughout the Solar System—are produced by collisions. An interloper smashes into a bigger world and makes an immense explosion at the point of impact. A bowl-shaped crater is excavated, and the smaller impacting object is destroyed. If the interlopers that dug out the Stickney and Herschel craters had been only a little larger, they would have had enough energy to blow Phobos and Mimas to bits. These moons barely escaped the cosmic wrecking ball. Many others did not.

Every time a world is smashed into, there's one less interloper—something like a demolition derby on the scale of the Solar System, a war of attrition. The very fact that many such collisions have occurred means that the rogue worldlets have been largely used up. Those on circular trajectories around the Sun, those that don't intersect the orbits of other worlds, will be unlikely to smash into a planet. Those on highly elliptical trajectories, those that cross the orbits of other planets, will sooner or later collide or, by a near miss, be gravitationally ejected from the Solar System.

The planets almost certainly accumulated from worldlets which in turn had condensed out of a great flat cloud of gas and dust surrounding the Sun—the sort of cloud that can now be seen around young nearby stars. So, in the early history of the Solar System before collisions cleaned things up, there should have been many more worldlets than we see today.

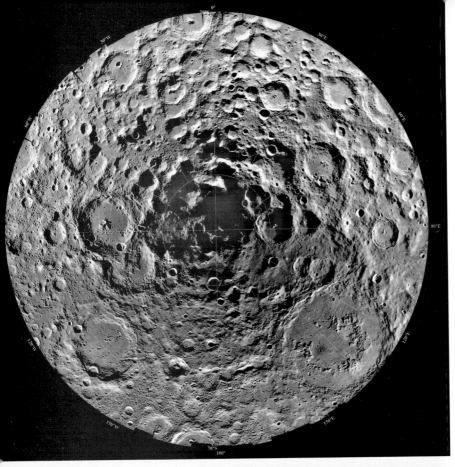

The battered and ravaged south polar region of the Moon. *Clementine* mission photomosaic, courtesy Naval Research Laboratory and USGS.

Cratered terrain on Mercury. Cratering is evident throughout the Solar System, from the innermost planet here to the moons of the outer planets. Violent collisions must have occurred routinely through the early history of the Solar System. *Mariner 10* photomosaic, courtesy USGS/NASA.

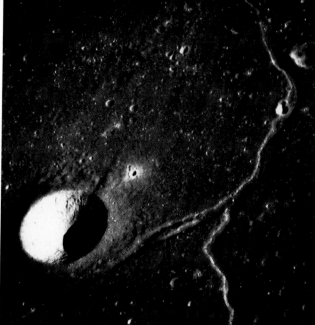

ABOVE LEFT: Impact craters exist here and there on the Earth's surface as well—but not nearly as many as on the Moon, because of the efficient erosion on our planet. This is an aerial photograph of Meteor Crater, Arizona, formed about 40,000 years ago. Copyright ©1994 by William K. Hartmann.

ABOVE RIGHT: Lunar basins flooded with lava have few craters—recording only those worldlets that impacted after the lava froze. Courtesy NASA

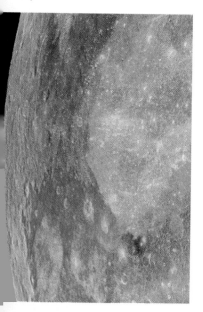

Craters dot the lunar landscape to the horizon in this false-color *Galileo* image. Courtesy JPL/NASA.

Indeed, there is clear evidence for this in our own backyard: If we count up the interloper worldlets in our neighborhood in space, we can estimate how often they'll hit the Moon. Let us make the very modest assumption that the population of interlopers has never been smaller than it is today. We can then calculate how many craters there should be on the Moon. The number we figure turns out to be much less than the number we see on the Moon's ravaged highlands. The unexpected profusion of craters on the Moon speaks to us of an earlier epoch when the Solar System was in wild turmoil, churning with worlds on collision trajectories. This makes good sense, because they formed from the aggregation of much smaller worldlets—which themselves had grown out of interstellar dust. Four billion years ago, the lunar impacts were hundreds of times more frequent than they are today; and 4.5 billion years ago, when the planets were still incomplete, collisions happened perhaps a billion times more often than in our becalmed epoch.

The chaos may have been relieved by much more flamboyant ring systems than grace the planets today. If they had small moons in that time, the Earth, Mars, and the other small planets may also have been adorned with rings.

The most satisfactory explanation of the origin of our own Moon, based on its chemistry (as revealed by samples returned from the *Apollo* missions), is that it was formed almost 4.5 billion years ago, when a world the size of Mars struck the Earth. Much of our planet's rocky mantle was reduced to dust and hot gas and blasted into space. Some of the debris, in orbit around the Earth, then gradually reaccumulated—atom by atom, boulder by boulder. If that unknown impacting world had been only a little larger, the result would have been the obliteration of the Earth. Perhaps there once were other worlds in our Solar System—perhaps even worlds on which life was stirring—hit by some demon worldlet, utterly demolished, and of which today we have not even an intimation.

The emerging picture of the early Solar System does not resemble a stately progression of events designed to form the Earth. Instead, it looks as if our planet was made, and survived, by mere lucky chance,* amid unbelievable violence. Our world does not seem to have been sculpted by a master craftsman. Here too, there is no hint of a Universe made for us.

* If it had not, perhaps there would today be another planet, a little nearer to or farther from the Sun, on which other, quite different beings would be trying to reconstruct *their* origins.

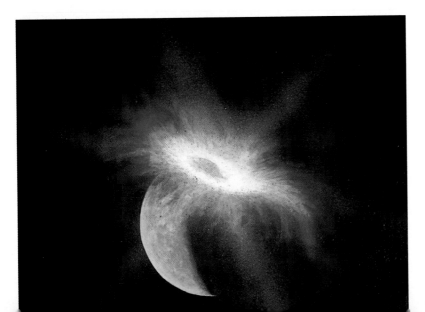

Perhaps the most massive impact ever to occur on Earth is thought to have happened about 4.4 billion years ago, forming the Moon. The impacting object was about the size of Mars. If it had been somewhat bigger, the Earth would have been destroyed. Painting by William K. Hartmann.

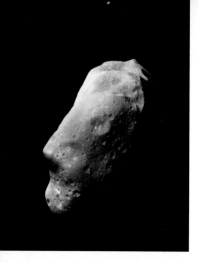

Main Belt asteroid 243 Ida, as imaged by *Galileo* on August 28, 1993. Ida is 52 kilometers long and rotates once every 4.6 hours. It is heavily cratered because of collisions with other, still smaller asteroids in the asteroid belt. Its moon is seen at top. Courtesy JPL/NASA.

The Main Belt asteroid Ida (above) displayed with the Martian moons Deimos (left) and Phobos (right), all to scale. Courtesy JPL/NASA.

THE DWINDLING SUPPLY of worldlets is today variously labeled: asteroids, comets, small moons. But these are arbitrary categories—real worldlets are able to breach these human-made partitions. Some asteroids (the word means "starlike," which they certainly are not) are rocky, others metallic, still others rich in organic matter. None is bigger than 1,000 kilometers across. They are found mainly in a belt between the orbits of Mars and Jupiter. Astronomers once thought the "main-belt" asteroids were the remains of a demolished world, but, as I've been describing, another idea is now more fashionable: The Solar System was once filled with asteroid-like worlds, some of which went into building the planets. Only in the asteroid belt, near Jupiter, did the gravitational tides of this most massive planet prevent the nearby debris from coalescing into a new world. The asteroids, instead of representing a world that once was, seem to be the building blocks of a world destined never to be.

Down to kilometer size, there may be several million asteroids, but, in the enormous volume of interplanetary space, even that's still far too few to cause any serious hazard to spacecraft on their way to the outer Solar System. The first main-belt asteroids, Gaspra and Ida, were photographed, in 1991 and 1993 respectively, by the *Galileo* spacecraft on its tortuous journey to Jupiter.

Main-belt asteroids mostly stay at home. To investigate them, we must go and visit them, as *Galileo* did. Comets, on the other hand, sometimes come and visit us, as Halley's comet did most recently in 1910 and 1986. Comets are made mainly of ice, plus smaller amounts of rocky and organic material. When heated, the ice vaporizes, forming the long and lovely tails blown outward by the solar wind and the pressure of sunlight. After many passages by the Sun, the ice is all evaporated, sometimes leaving a dead rocky and organic world. Sometimes the remaining particles, the ice that held them together now gone, spread out in the comet's orbit, generating a debris trail around the Sun.

Every time a bit of cometary fluff the size of a grain of sand enters the Earth's atmosphere at high speed, it burns up, producing a momentary trail of light that Earthbound observers call a sporadic meteor or "shooting star." Some disintegrating comets have orbits that cross the Earth's. So every year, the Earth, on its steady circumnavigation of the Sun, also plunges through belts of orbit-

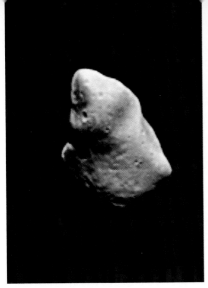

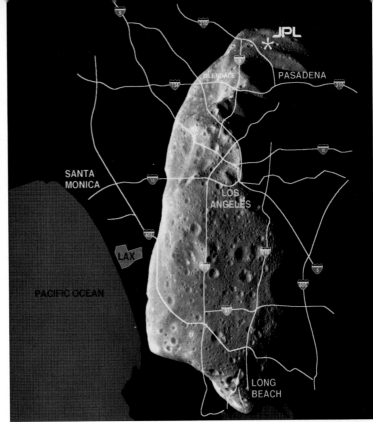

ABOVE LEFT: *Galileo* image of the Main Belt asteroid 951 Gaspra, visited during *Galileo's* long, looping trajectory to Jupiter. Courtesy JPL/NASA.

ABOVE RIGHT: Gaspra compared with the Los Angeles freeway system. Courtesy JPL/NASA.

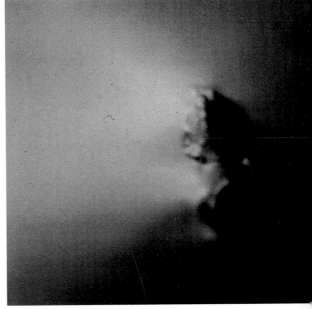

Two views of the nucleus of Halley's Comet. The nucleus is very dark and covered with organic matter. Jets of water vapor and fine particles are streaming off its surface, where they will be molded by the pressure of sunlight and the solar wind into a glorious tail. Halley's Comet is also about 10 kilometers across, the size of the Cretaceous-Tertiary impactor. Images from the Halley Multicolour Camera aboard the *Giotto* spacecraft of the European Space Agency. Courtesy ESA.

ing cometary debris. We may then witness a meteor shower, or even a meteor storm—the skies ablaze with the body parts of a comet. For example, the Perseid meteors, seen on or about August 12 of each year, originate in a dying comet called Swift-Tuttle. But the beauty of a meteor shower should not deceive us: There is a continuum that connects these shimmering visitors to our night skies with the destruction of worlds.

A few asteroids now and then give off little puffs of gas or even form a temporary tail, suggesting that they are in transition between cometdom and asteroidhood. Some small moons going around the planets are probably captured asteroids or comets; the moons of Mars and the outer satellites of Jupiter may be in this category.

Gravity smooths down everything that sticks out too far. But only in large bodies is the gravity enough to make mountains and other projections collapse of their own weight, rounding the world. And, indeed, when we observe their shapes, almost always we find that small worldlets are lumpy, irregular, potato-shaped.

THERE ARE ASTRONOMERS whose idea of a good time is to stay up till dawn on a cold, moonless night taking pictures of the sky —the same sky they photographed the year before . . . and the year before that. If they got it right last time, you might well ask, why are they doing it again? The answer is: The sky changes. In any given year there might be worldlets wholly unknown, never seen before, that approach the Earth and are spied by these dedicated observers.

On March 25, 1993, a group of asteroid and comet hunters, looking at the photographic harvest from an intermittently cloudy night at Mount Palomar in California, discovered a faint elongated smudge on their films. It was near a very bright object in the sky, the planet Jupiter. Carolyn and Eugene Shoemaker and David Levy then asked other observers to take a look. The smudge turned out to be something astonishing: some twenty small, bright objects orbiting Jupiter, one behind the other, like pearls on a string. Collectively they are called Comet Shoemaker-Levy 9 (this is the ninth time that these collaborators have together discovered a periodic comet).

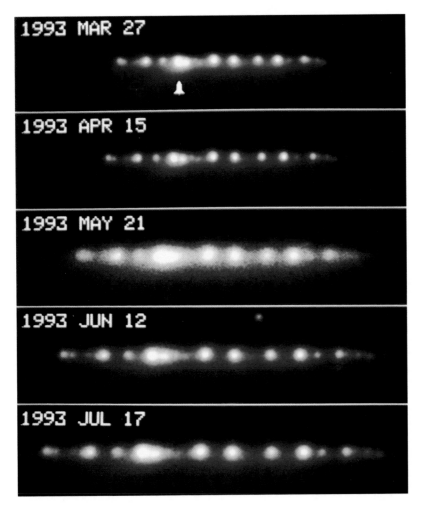

1993 MAR 27

1993 APR 15

1993 MAY 21

1993 JUN 12

1993 JUL 17

Observations of the components of comet Shoemaker-Levy 9 as observed in spring and summer 1993. As time goes on, the fragments, although confined to the same orbit, separate from one another. Courtesy David Jewett, University of Hawaii.

But calling these objects *a* comet is confusing. There was a horde of them, probably the fragmented remains of a single, hitherto undiscovered, comet. It silently orbited the Sun for 4 billion years before passing too close to Jupiter and being captured, a few decades ago, by the gravity of the Solar System's largest planet. On July 7, 1992, it was torn apart by Jupiter's gravitational tides.

You can recognize that the inner part of such a comet would be pulled toward Jupiter a little more strongly than the outer part, because the inner part is closer to Jupiter than the outer part. The difference in pull is certainly small. Our feet are a little closer to the center of the Earth than our heads, but we are not in consequence torn to pieces by the Earth's gravity. For such tidal disruption to have occurred, the original comet must have been held together very weakly. Before fragmentation, it was, we think, a loosely consolidated mass of ice, rock, and organic matter, maybe 10 kilometers (about 6 miles) across.

The orbit of this disrupted comet was then determined to high precision. Between July 16 and 22, 1994, all the cometary fragments, one after another, collided with Jupiter. The biggest pieces seem to have been a few kilometers across. Their impacts with Jupiter were spectacular.

No one knew beforehand what these multiple impacts into the atmosphere and clouds of Jupiter would do. Perhaps the cometary fragments, surrounded by halos of dust, were much smaller than they seemed. Or perhaps they were not coherent bodies at all, but loosely consolidated—something like a heap of gravel with all the particles traveling through space together, in nearly identical orbits. If either of these possibilities were true, Jupiter might swallow the comets without a trace. Other astronomers thought there would at least be bright fireballs and giant plumes as the cometary fragments plunged into the atmosphere. Still others suggested that the dense cloud of fine particles accompanying the fragments of Comet Shoemaker-Levy 9 into Jupiter would disrupt the magnetosphere of Jupiter or form a new ring.

A comet this size should impact Jupiter, it is calculated, only once every thousand years. It's the astronomical event not of one lifetime, but of a dozen. Nothing on this scale has occurred since the invention of the telescope. So in mid-July 1994, in a beautifully coordinated international scientific effort, telescopes all over the Earth and in space turned towards Jupiter.

Astronomers had over a year to prepare. The trajectories of the fragments in their orbits around Jupiter were estimated. It was discovered that they would all hit Jupiter. Predictions of the timing were refined. Disappointingly, the calculations revealed that all impacts would occur on the night side of Jupiter, the side invisible from the Earth (although accessible to the *Galileo* and *Voyager* spacecraft in the outer Solar System). But, happily, all impacts

The fragments of comet Shoemaker-Levy 9, as seen from the Hubble Space Telescope, embedded in a dust cloud shed by the comets. Courtesy H. A. Weaver and T. E. Smith, Space Telescope Science Institute/NASA.

would occur only a few minutes before the Jovian dawn, before the impact site would be carried by Jupiter's rotation into the line of sight from Earth.

The appointed moment for the impact of the first piece, Fragment A, came and went. There were no reports from ground-based telescopes. Planetary scientists stared with increasing gloom at a television monitor displaying the data transmitted to the Space Telescope Science Institute in Baltimore from the Hubble Space Telescope. There was nothing anomalous. Shuttle astronauts took time off from the reproduction of fruit flies, fish, and newts to look at Jupiter through binoculars. They reported seeing nothing. The impact of the millennium was beginning to look very much like a fizzle.

Then there was a report from a ground-based optical telescope in La Palma in the Canary Islands, followed by announcements from a radiotelescope in Japan; from the European Southern Observatory in Chile; and from a University of Chicago instrument in the frigid wastelands of the South Pole. In Baltimore the young scientists crowding around the TV monitor—themselves monitored by the cameras of CNN—began to see something, and in exactly the right place on Jupiter. You could witness consternation turn into puzzlement, and then exultation. They cheered; they screamed; they jumped up and down. Smiles filled the room. They broke out the champagne. Here was a group of young American scientists—about a third of them, including the team leader, Heidi Hammel, women—and you could imagine youngsters all over the world thinking that it might be fun to be a scientist, that this might be a good daytime job, or even a means to spiritual fulfillment.

For many of the fragments, observers somewhere on Earth noticed the fireball rise so quickly and so high that it could be seen even though the impact site below it was still in Jovian darkness. Plumes ascended and then flattened into pancake-like forms. Spreading out from the point of impact we could see sound and gravity waves, and a patch of discoloration that for the largest fragments became as big as the Earth.

Slamming into Jupiter at 60 kilometers a second (130,000 miles an hour), the large fragments converted their kinetic energy partly into shock waves, partly into heat. The temperature in the fireball was estimated at thousands of degrees. Some of the fireballs and plumes were far brighter than all the rest of Jupiter put together.

What is the cause of the dark stains left after the impact? It might be stuff from the deep clouds of Jupiter—from the region to which ground-based observers cannot ordinarily see—that welled up and spread out. However, the fragments do not seem to have penetrated to such depths. Or the molecules responsible for the stains might have been in the cometary fragments in the first place. We know from the *Vega 1* and *2* Soviet missions and the

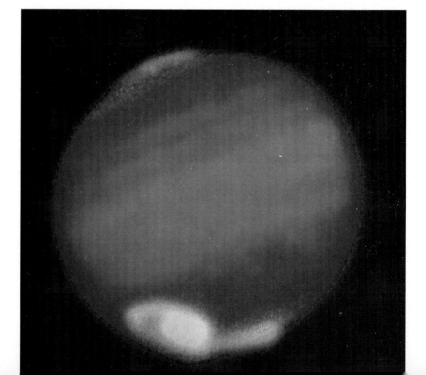

The largest piece, Fragment G, of Comet Shoemaker-Levy 9 impacts Jupiter. The plume and its surrounding ring of hot gas is seen in light blue in this infrared false-color image. Courtesy Peter McGregor and Mark Allen, Australian National University Telescope at Siding Spring.

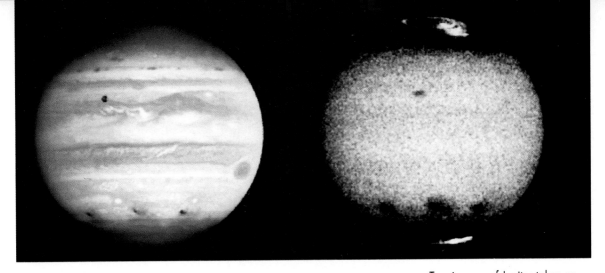

Giotto mission of the European Space Agency—both to Halley's Comet—that comets may be as much as a quarter composed of complex organic molecules. They are the reason that the nucleus of Halley's Comet is pitch black. If some of the cometary organics survived the impact events, they may have been responsible for the stain. Or, finally, the stain may be due to organic matter not delivered by the impacting cometary fragments, but synthesized by their shock waves from the atmosphere of Jupiter.

Impact of the fragments of Comet Shoemaker-Levy 9 with Jupiter was witnessed on seven continents. Even amateur astronomers with small telescopes could see the plumes and the subsequent discoloration of the Jovian clouds. Just as sporting events are covered at all angles by television cameras on the field and from a dirigible high overhead, six NASA spacecraft deployed throughout the Solar System, with different observational specialties, recorded this new wonder—the *Hubble Space Telescope*, the *International Ultraviolet Explorer*, and the *Extreme Ultraviolet Explorer* all in Earth orbit; *Ulysses*, taking time out from its investigation of the South Pole of the Sun; *Galileo*, on the way to its own rendezvous with Jupiter; and *Voyager 2*, far beyond Neptune on its way to the stars. As the data are accumulated and analyzed, our knowledge of comets, of Jupiter, and of the violent collisions of worlds should all be substantially improved.

For many scientists—but especially for Carolyn and Eugene Shoemaker and David Levy—there was something poignant about the cometary fragments, one after the other, making their death plunges into Jupiter. They had lived with this comet, in a manner of speaking, for 16 months, watched it split, the pieces, enshrouded

Two images of Jupiter taken on July 17, 1994—at left in violet light and at right in ultraviolet light. The three dark spots in the southern hemisphere are the collision sites of Fragments C, A, and E, counting left to right, of Shoemaker-Levy 9. In violet light, these three dark spots are approximately as large as the Earth. They are considerably bigger in ultraviolet light. The spots may be due to complex organic molecules carried to Jupiter from the comet or generated in the Jovian atmosphere by the comets's shock waves. Note the dark violet and ultraviolet polar caps, probably caused by complex organics generated by electrons pouring into the atmosphere near the poles from Jupiter's magnetosphere. The bright traceries at the poles are Jovian auroras. Images obtained by the Wide Field Planetary Camera 2 of the Hubble Space Telescope. Courtesy John Clarke, University of Michigan and NASA.

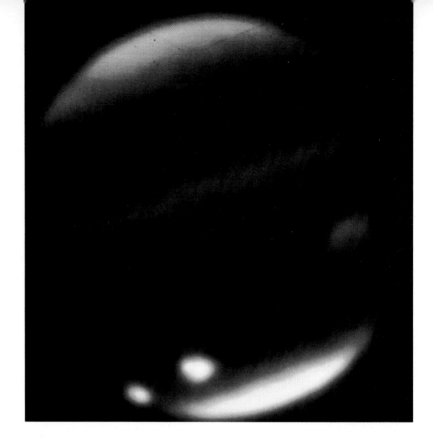

Impacts of Fragments A and C
of Comet Shoemaker-Levy 9
on Jupiter (bottom left).
This picture was taken by the
Keck Observatory in Hawaii,
the world's largest
optical telescope.

A component of Comet
Shoemaker-Levy 9 impacts the
atmosphere of Jupiter, churning
up a fireball from the depths.
Painting by Don Davis.

by clouds of dust, playing hide-and-seek and spreading out in their orbits. In a limited way, each fragment had its own personality. Now they're all gone, ablated into molecules and atoms in the upper atmosphere of the Solar System's largest planet. In a way, we almost mourn them. But we're learning from their fiery deaths. It is perhaps some reassurance to know that there are a hundred trillion more of them in the Sun's vast treasure-house of worlds.

THERE ARE ABOUT 200 known asteroids whose paths take them near the Earth. They are called, appropriately enough, "near-Earth" asteroids. Their detailed appearance (like that of their main-belt cousins) immediately implies that they are the products of a violent collisional history. Many of them may be the shards and remnants of once-larger worldlets.

With a few exceptions, the near-Earth asteroids are only a few kilometers across or smaller, and take one to a few years to make a circuit around the Sun. About 20 percent of them, sooner or later, are bound to hit the Earth—with devastating consequences. (But in astronomy, "sooner or later" can encompass billions of years.) Cicero's assurance that "nothing of chance or

SOME NEAR-EARTH ASTEROID ORBITS

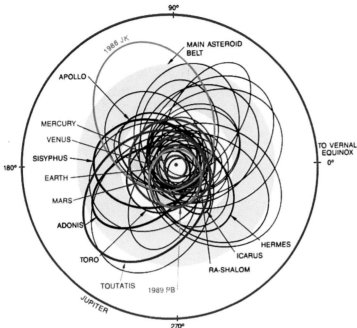

A sampling of a few of the estimated 2,000 largish asteroids whose orbits cross the Earth's. Orbits of Mercury, Venus, Earth, Mars, and Jupiter are shown in red. Sooner or later, some of these worldlets are bound to hit the Earth. Diagram courtesy JPL/NASA.

hazard" is to be found in an absolutely ordered and regular heaven is a profound misperception. Even today, as Comet Shoemaker-Levy's encounter with Jupiter reminds us, there is routine interplanetary violence, although not on the scale that marked the early history of the Solar System.

Like main-belt asteroids, many near-Earth asteroids are rocky. A few are mainly metal, and it has been suggested that enormous rewards might attend moving such an asteroid into orbit around the Earth, and then systematically mining it—a mountain of high-grade ore a few hundred miles overhead. The value of platinum-group metals alone in a single such world has been estimated as many trillions of dollars—although the unit price would plummet spectacularly if such materials became widely available. Methods of extracting metals and minerals from appropriate asteroids are being studied, for example by John Lewis, a planetary scientist at the University of Arizona.

Some near-Earth asteroids are rich in organic matter, apparently preserved from the very earliest Solar System. Some have

Unidentified flying molars: Shown here is a computer model of the near-Earth asteroid 4769 Castalia rotating about a vertical axis. The model is based on Arecibo radar data obtained by Steven Ostro of JPL and his colleagues in 1989, when the asteroid was 5.6 million kilometers (3.5 million miles) away. At higher resolution this double world would doubtless show many craters. It may have been formed when one asteroid gently nudged into another, a process that may shed light on the origin of planets. Courtesy JPL/NASA.

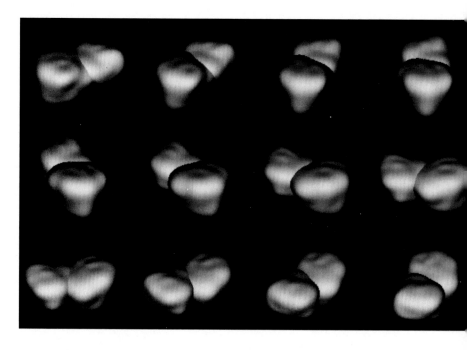

been found, by Steven Ostro of the Jet Propulsion Laboratory, to be double, two bodies in contact. Perhaps a larger world has broken in two as it passed through the strong gravitational tides of a planet like Jupiter; more interesting is the possibility that two worlds on similar orbits made a gentle overtaking collision and stuck. This process may have been key to the building of planets and the Earth. At least one asteroid (Ida, as viewed by *Galileo*) has its own small moon. We might guess that two asteroids in contact and two asteroids orbiting one another have related origins.

Sometimes, we hear about an asteroid making a "near miss." (Why do we call it a "near miss"? A "near hit" is what we really mean.) But then we read a little more carefully, and it turns out that its closest approach to the Earth was several hundreds of thousands or millions of kilometers. That doesn't count—that's too far away, farther even than the Moon. If we had an inventory of all the near-Earth asteroids, including those considerably smaller than a kilometer across, we could project their orbits into the future and predict which ones are potentially dangerous. There are an estimated 2,000 of them bigger than a kilometer across, of which we have actually observed only a few percent. There are maybe 200,000 bigger than 100 meters in diameter.

The near-Earth asteroids have evocative mythological names: Orpheus, Hathor, Icarus, Adonis, Apollo, Cerberus, Khufu, Amor, Tantalus, Aten, Midas, Ra-Shalom, Phaethon, Toutatis, Quetzal-coatl. There are a few of special exploratory potential—for example, Nereus. In general, it's much easier to get onto and off of near-Earth asteroids than the Moon. Nereus, a tiny world about a kilometer across, is one of the easiest.* It would be real exploration of a truly new world.

Some humans (all from the former Soviet Union) have already been in space for periods longer than the entire round-trip time to Nereus. The rocket technology to get there already exists. It's a much smaller step than going to Mars or even, in several respects, than returning to the Moon. If something went wrong, though, we would be unable to run home to safety in only a few days. In this respect, its level of difficulty lies somewhere between a voyage to Mars and one to the Moon.

Of many possible future missions to Nereus, there's one that takes 10 months to get there from Earth, spends 30 days there, and then requires only 3 weeks to return home. We could visit Nereus with robots, or—if we're up to it—with humans. We could examine this little world's shape, constitution, interior, past history, organic chemistry, cosmic evolution, and possible tie to comets. We could bring samples back for examination at leisure in Earth-bound laboratories. We could investigate whether there really are commercially valuable resources—metals or minerals—there. If we are ever going to send humans to Mars, near-Earth asteroids provide a convenient and appropriate intermediate goal—to test out the equipment and exploratory protocols while studying an almost wholly unknown little world. Here's a way to get our feet wet again when we're ready to re-enter the cosmic ocean.

---

* Asteroid 1991JW has an orbit very much like the Earth's and is even easier to get to than 4660 Nereus. But its orbit seems *too* similar to the Earth's for it to be a natural object. Perhaps it's some lost upper stage of the *Saturn V Apollo* Moon rocket.

THE MARSH OF CAMARINA

[I]t's too late to make any improvements now. The universe is finished;
the copestone is on, and the chips were carted off a million years ago.

—HERMAN MELVILLE, *MOBY DICK*, CHAPTER 2 (1851)

Camarina was a city in southern Sicily, founded by colonists from Syracuse in 598 B.C. A generation or two later, it was threatened by a pestilence—festering, some said, in the adjacent marsh. (While the germ theory of disease was certainly not widely accepted in the ancient world, there were hints—for example, Marcus Varro in the first century B.C. advised explicitly against building cities near swamps "because there are bred certain minute creatures which cannot be seen by the eyes, which float in the air and enter the body through the mouth and nose and there cause serious disease.") The danger to Camarina

OPPOSITE: The Cretaceous epoch of geological time ends: A 10-kilometer-diameter asteroid or comet slams into the Earth near what today is the Yucatán Peninsula. All the dinosaurs and 75 percent of all the other species of life on Earth will be extinguished as a result. Painting by Don Davis.

An interplanetary object, probably a fragment of a comet, sweeps through the Earth's atmosphere and dissipates before striking the surface. (Many stars can be seen. By accident, the fireball passes directly in front of a distant spiral galaxy.) Impacting objects in the several-hundred-meter size range and larger threaten the global civilization. Courtesy Anglo-Australian Observatory. Photograph by David Malin.

was great. Plans were drawn to drain the marsh. When the oracle was consulted, though, it forbade such a course of action, counseling patience instead. But lives were at stake, the oracle was ignored, and the marsh was drained. The pestilence was promptly halted. Too late, it was recognized that the marsh had protected the city from its enemies—among whom there had now to be counted their cousins the Syracusans. As in America 2,300 years later, the colonists had quarreled with the mother country. In 552 B.C., a Syracusan force crossed over the dry land where the marsh had been, slaughtered every man, woman, and child, and razed the city. The marsh of Camarina became proverbial for eliminating a danger in such a way as to usher in another, much worse.

THE CRETACEOUS-TERTIARY COLLISION (or collisions—there may have been more than one) illuminates the peril from asteroids and comets. In sequence, a world-immolating fire burned vegetation to a crisp all over the planet; a stratospheric dust cloud so darkened the sky that surviving plants had trouble making a living from photosynthesis; there were worldwide freezing temperatures, torrential rains of caustic acids, massive depletion of the ozone layer, and, to top it off, after the Earth healed itself from these assaults, a prolonged greenhouse warming (because the main impact seems to have volatilized a deep layer of sedimentary carbonates, pouring huge amounts of carbon dioxide into the air). It was not a single catastrophe, but a parade of them, a concatenation of terrors. Organisms weakened by one disaster were finished off by the next. It is quite uncertain whether our civilization would survive even a considerably less energetic collision.

Since there are many more small asteroids than large ones, run-of-the-mill collisions with the Earth will be made by the little guys. But the longer you're prepared to wait, the more devastating the impact you can expect. On average, once every few hundred years the Earth is hit by an object about 70 meters in diameter; the resulting energy released is equivalent to the largest nuclear weapons explosion ever detonated. Every 10,000 years, we're hit by a 200-meter object that might induce serious regional climatic effects. Every million years, an impact by a body over 2 kilometers in diameter occurs, equivalent to nearly a million megatons of TNT—an explosion that would work a global catastrophe, killing (unless unprecedented precautions were taken) a significant fraction of the human species. A million megatons of TNT is 100 times the explosive yield of all the nuclear weapons on the planet, if simultaneously blown up. Dwarfing even this, in a hundred million years or so, you can bet on something like the Cretaceous-Tertiary event, the impact of a world 10 kilometers across or bigger. The destructive energy latent in a large near-Earth asteroid dwarfs anything else the human species can get its hands on.

As first shown by the American planetary scientist Christopher Chyba and his colleagues, little asteroids or comets, a few tens of meters across, break and burn up on entering our atmosphere. They arrive comparatively often but do no significant harm. Some idea of how frequently they enter the Earth's atmosphere has been revealed by declassified Department of Defense data obtained from special satellites monitoring the Earth for clandestine nuclear explosions. There seem to have been hundreds of small worldlets

The Earth shortly after the impact of a 10-kilometer asteroid or comet like that which extinguished most species of life 65 million years ago. In this artist's conception, the impact occurs on the East Coast of the United States near Washington, D.C. The impact crater is beginning to fill with water from the Chesapeake Bay. Fires are set all over Earth. Painting by Don Davis.

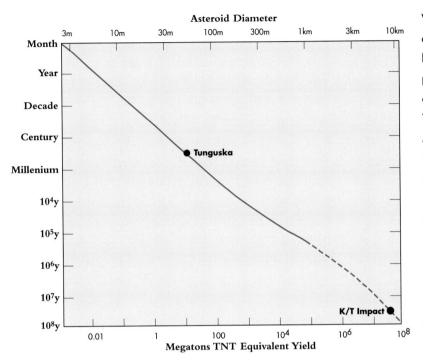

What size asteroids do how much damage, and how often do they hit the Earth? This diagram, prepared by Clark R. Chapman of the Planetary Sciences Institute, Tucson, Arizona, and David Morrison of the NASA Ames Research Center, summarizes the best of current knowledge. You read it this way: Consider the point marked "Tunguska," an object that entered the Earth's atmosphere over Siberia in the year 1908. While it dissipated before digging out a crater in the ground, it was powerful enough to knock down forests and to be detected halfway around the world. An event like Tunguska would be caused by an asteroid about 50 meters in diameter (scale at top), and would release an energy equivalent to about 10 megatons of TNT (scale at bottom)—a hefty, but not the most powerful, contemporary nuclear weapon. Reading off the vertical axis, we see that we should expect a Tunguska-scale impact once every few centuries. As we walk down the curve to the right, we come to larger bodies, more dangerous impacts, and a longer waiting time for such impacts to occur. The Cretaceous-Tertiary (K/T) impact is shown at bottom right.

(and at least one larger body) impacting in the last 20 years. They did no harm. But, we need to be very sure we can distinguish a small colliding comet or asteroid from an atmospheric nuclear explosion.

Civilization-threatening impacts require bodies several hundred meters across, or more. (A meter is about a yard; 100 meters is roughly the length of a football field.) They arrive something like once every 200,000 years. Our civilization is only about 10,000 years old, so we should have no institutional memory of the last such impact. Nor do we.

Comet Shoemaker-Levy 9, in its succession of fiery explosions on Jupiter in July 1994, reminds us that such impacts really do occur in our time—and that the impact of a body a few kilometers across can spread debris over an area as big as the Earth. It was a kind of portent.

In the very week of the Shoemaker-Levy impact, the Science and Space Committee of the U.S. House of Representatives drafted legislation that requires NASA "in coordination with the Department of Defense and the space agencies of other countries" to identify and determine the orbital characteristics of all Earth-approaching "comets and asteroids that are greater than 1 kilometer in diameter." The work is to be completed by the year 2005. Such a search program had been advocated by many planetary scientists. But it took the death throes of a comet to move it toward practical implementation.

Spread out over the waiting time, the dangers of asteroid collision do not seem very worrisome. But if a big impact happens, it would be an unprecedented human catastrophe. There's something like one chance in two thousand that such a collision will happen in the lifetime of a newborn baby. Most of us would not fly in an airplane if the chance of crashing were one in two thousand. (In fact for commercial flights the chance is one in two million. Even so, many people consider this large enough to worry about, or even to take out insurance for.) When our lives are at stake, we often change our behavior to arrange more favorable odds. Those who don't tend to be no longer with us.

Perhaps we should practice getting to these worldlets and diverting their orbits, should the hour of need ever arise. Melville notwithstanding, some of the chips of creation *are* still left, and improvements evidently need to be made. Along parallel and only weakly interacting tracks, the planetary science community and the U.S. and Russian nuclear weapons laboratories, aware of the foregoing scenarios, have been pursuing these questions: how to monitor all sizable near-Earth interplanetary objects, how to characterize their physical and chemical nature, how to predict which ones may be on a future collision trajectory with Earth, and, finally, how to prevent a collision from happening.

The Russian spaceflight pioneer Konstantin Tsiolkovsky argued a century ago that there must be bodies intermediate in size between the observed large asteroids and those asteroidal fragments, the meteorites, that occasionally fall to Earth. He wrote about living on small asteroids in interplanetary space. He did not have military applications in mind. In the early 1980s, though, some in the U.S. weapons establishment argued that the Soviets might use near-Earth asteroids as first-strike weapons; the alleged plan was called "Ivan's Hammer." Countermeasures were needed.

Looking something like a black eye is the discoloration in the Jupiter clouds produced by Fragment G of Comet Shoemaker-Levy 9 on July 18, 1994. The larger, very dark oval is about the size of the Earth. It is surrounded by a spreading sound wave, outside of which is a fainter discoloration. The smaller dark spot is the impact scar of Fragment D. This picture is a useful reminder that a comet or asteroid a few kilometers in size can generate debris over an area the size of the Earth. Hubble Space Telescope image, courtesy Heidi Hammel, MIT, and NASA.

But, at the same time, it was suggested, maybe it wasn't a bad idea for the United States to learn how to use small worlds as weapons of its own. The Defense Department's Ballistic Missile Defense Organization, the successor to the Star Wars office of the 1980s, launched an innovative spacecraft called *Clementine* to orbit the Moon and fly by the near-Earth asteroid Geographos. (After completing a remarkable reconnaissance of the Moon in May 1994, the spacecraft failed before it could reach Geographos.)

Asteroid deflection: A dangerous asteroid—moving (in stopped motion) along its orbit away from the reader in left foreground—would, if unhindered, impact the Earth months later. Instead, at the proper moment one or more missiles are launched from Earth (orbit in red) to explode nuclear weapons in the vicinity of the closest point in the asteroid's orbit to the Sun. A relatively small nudge is sufficient to alter the orbit (lavender) so it misses the Earth. Diagram courtesy JPL/NASA.

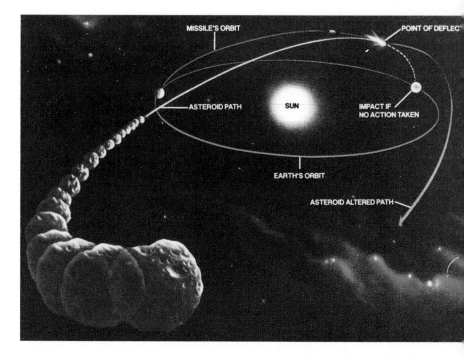

In principle, you could use big rocket engines, or projectile impact, or equip the asteroid with giant reflective panels and shove it with sunlight or powerful Earth-based lasers. But with technology that exists right now, there are only two ways. First, one or more high-yield nuclear weapons might blast the asteroid or comet into fragments that would disintegrate and atomize on entering the Earth's atmosphere. If the offending worldlet is only weakly held together, perhaps only hundreds of megatons would suffice. Since there is no theoretical upper limit to the explosive yield of a thermonuclear weapon, there seem to be those in the weapons laboratories who consider making bigger bombs not only as a stirring challenge, but also as a way to mute pesky environmentalists by securing a seat for nuclear weapons on the save-the-Earth bandwagon.

Another approach under more serious discussion is less dramatic but still an effective way of maintaining the weapons establishment—a plan to alter the orbit of any errant worldlet by exploding nuclear weapons nearby. The explosions (generally near the asteroid's closest point to the Sun) are arranged to deflect it away from the Earth.* A flurry of low-yield nuclear weapons, each giving a little push in the desired direction, is enough to deflect a medium-sized asteroid with only a few weeks' warning. The method also offers, it is hoped, a way to deal with a suddenly detected long-period comet on imminent collision trajectory with the Earth: The comet would be intercepted with a small asteroid. (Needless to say, this game of celestial billiards is even more difficult and uncertain—and therefore even less practical in the near future—than the herding of an asteroid on a known, well-behaved orbit with months or years at our disposal.)

We don't know what a standoff nuclear explosion would do to an asteroid. The answer may vary from asteroid to asteroid. Some small worlds might be strongly held together; others might be little more than self-gravitating gravel heaps. If an explosion breaks, let's say, a 10-kilometer asteroid up into hundreds of 1-kilometer fragments, the likelihood that at least one of them impacts the Earth is probably increased, and the apocalyptic character of the consequences may not be much reduced. On the other hand, if the explosion disrupts the asteroid into a swarm of objects a hundred meters in diameter or smaller, all of them might ablate away like giant meteors on entering the Earth's atmosphere. In this case little impact damage would be caused. Even if the asteroid were wholly pulverized into fine powder, though, the resulting high-altitude dust layer might be so opaque as to block the sunlight and change the climate. We do not yet know.

A vision of dozens or hundreds of nuclear-armed missiles on ready standby to deal with threatening asteroids or comets has

---

* The Outer Space Treaty, adhered to both by the United States and Russia, prohibits weapons of mass destruction in "outer space." Asteroid deflection technology constitutes just such a weapon—indeed, the most powerful weapon of mass destruction ever devised. Those interested in developing asteroid deflection technology will want to have the treaty revised. But even with no revision, were a large asteroid to be discovered on impact trajectory with the Earth, presumably no one's hand would be stayed by the niceties of international diplomacy. There is a danger, though, that relaxing prohibitions on such weapons in space might make us less attentive about the positioning of warheads for offensive purposes in space.

been offered. However premature in this particular application, it seems very familiar; only the enemy has been changed. It also seems very dangerous.

The problem, Steven Ostro of JPL and I have suggested, is that if you can reliably deflect a threatening worldlet so it does not collide with the Earth, you can also reliably deflect a harmless worldlet so it *does* collide with the Earth. Suppose you had a full inventory, with orbits, of the estimated 300,000 near-Earth asteroids larger than 100 meters—each of them large enough, on impacting the Earth, to have serious consequences. Then, it turns out, you also have a list of huge numbers of inoffensive asteroids whose orbits could be altered with nuclear warheads so they quickly collide with the Earth.

Suppose we restrict our attention to the 2,000 or so near-Earth asteroids that are a kilometer across or bigger—that is, the ones most likely to cause a global catastrophe. Today, with only about 100 of these objects catalogued, it would take about a century to catch one when it's easily deflectable to Earth and alter its orbit. We think we've found one, an as-yet-unnamed[*] asteroid so far denoted only as 1991OA. In 2070, this world, about 1 kilometer in diameter, will come within 4.5 million kilometers of the Earth's orbit—only fifteen times the distance to the Moon. To deflect 1991OA so it hits the Earth, only about 60 megatons of TNT equivalent needs to be exploded in the right way—the equivalent of a small number of currently available nuclear warheads.

Now imagine a time, a few decades hence, when all such near-Earth asteroids are inventoried and their orbits compiled. Then, as Alan Harris of JPL, Greg Canavan of the Los Alamos National Laboratory, Ostro, and I have shown, it might take only a year to select a suitable object, alter its orbit, and send it crashing into the Earth with cataclysmic effect.

The technology required—large optical telescopes, sensitive detectors, rocket propulsion systems able to lift a few tons of payload and make precise rendezvous in nearby space, and thermonu-

---

[*] What should we call this world? Naming it after the Greek Fates or Furies or Nemesis seems inappropriate, because whether it misses or hits the Earth is entirely in our hands. If we leave it alone, it misses. If we push it cleverly and precisely, it hits. Maybe we should call it "Eight Ball."

clear weapons—all exist today. Improvements in all but perhaps the last can be confidently expected. If we're not careful, many nations may have these capabilities in the next few decades. What kind of world will we then have made?

We have a tendency to minimize the dangers of new technologies. A year before the Chernobyl disaster, a Soviet nuclear power industry deputy minister was asked about the safety of Soviet reactors, and chose Chernobyl as a particularly safe site. The average waiting time to disaster, he confidently estimated, was a hundred thousand years. Less than a year later . . . devastation. Similar reassurances were provided by NASA contractors the year before the *Challenger* disaster: You would have to wait ten thousand years, they estimated, for a catastrophic failure of the shuttle. One year later . . . heartbreak.

Chlorofluorocarbons (CFCs) were developed specifically as a completely safe refrigerant—to replace ammonia and other refrigerants that, on leaking out, had caused illness and some deaths. Chemically inert, nontoxic (in ordinary concentrations), odorless, tasteless, nonallergenic, nonflammable, CFCs represent a brilliant technical solution to a well-defined practical problem. They found uses in many other industries besides refrigeration and air conditioning. But, as I described above, the chemists who developed CFCs overlooked one essential fact—that the molecules' very inertness guarantees that they are circulated to stratospheric altitudes and there cracked open by sunlight, releasing chlorine atoms which then attack the protective ozone layer. Due to the work of a few scientists, the dangers may have been recognized and averted in time. We humans have now almost stopped producing CFCs. We won't actually know if we've avoided real harm for about a century; that's how long it takes for all the CFC damage to be completed. Like the ancient Camarinans, we make mistakes.* Not only do we often ignore the warnings of the oracles; characteristically we do not even consult them.

---

* There is of course a wide range of other problems brought on by the devastatingly powerful technology we've recently invented. But in most cases they're not Camarinan disasters—damned if you do and damned if you don't. Instead they're dilemmas of wisdom or timing—for example, the wrong refrigerant or refrigeration physics out of many possible alternatives.

The notion of moving asteroids into Earth orbit has proved attractive to some space scientists and long-range planners. They foresee mining the minerals and precious metals of these worlds or providing resources for the construction of space infrastructure without having to fight the Earth's gravity to get them up there. Articles have been published on how to accomplish this end and what the benefits will be. In modern discussions, the asteroid is inserted into orbit around the Earth by first making it pass through and be braked by the Earth's atmosphere, a maneuver with very little margin for error. For the near future we can, I think, recognize this whole endeavor as unusually dangerous and foolhardy, especially for metal worldlets larger than tens of meters across. This is the one activity where errors in navigation or propulsion or mission design can have the most sweeping and catastrophic consequences.

The foregoing are examples of inadvertence. But there's another kind of peril: We are sometimes told that this or that invention would of course not be misused. No sane person would be so reckless. This is the "only a madman" argument. Whenever I hear it (and it's often trotted out in such debates), I remind myself that madmen really exist. Sometimes they achieve the highest levels of political power in modern industrial nations. This is the century of Hitler and Stalin, tyrants who posed the gravest dangers not just to the rest of the human family, but to their own people as well. In the winter and spring of 1945, Hitler ordered Germany to be destroyed—even "what the people need for elementary survival"—because the surviving Germans had "betrayed" him, and at any rate were "inferior" to those who had already died. If Hitler had had nuclear weapons, the threat of a counterstrike by Allied nuclear weapons, had there been any, is unlikely to have dissuaded him. It might have encouraged him.

Can we humans be trusted with civilization-threatening technologies? If the chance is almost one in a thousand that much of the human population will be killed by an impact in the next century, isn't it more likely that asteroid deflection technology will get into the wrong hands in another century—some misanthropic sociopath like a Hitler or a Stalin eager to kill everybody, a megalomaniac lusting after "greatness" and "glory," a victim of ethnic

violence bent on revenge, someone in the grip of unusually severe testosterone poisoning, some religious fanatic hastening the Day of Judgment, or just technicians incompetent or insufficiently vigilant in handling the controls and safeguards? Such people exist. The risks seem far worse than the benefits, the cure worse than the disease. The cloud of near-Earth asteroids through which the Earth plows may constitute a modern Camarine marsh.

It's easy to think that all of this must be very unlikely, mere anxious fantasy. Surely sober heads would prevail. Think of how many people would be involved in preparing and launching warheads, in space navigation, in detonating warheads, in checking what orbital perturbation each nuclear explosion has made, in herding the asteroid so it is on an impact trajectory with Earth, and so on. Isn't it noteworthy that although Hitler gave orders for the retreating Nazi troops to burn Paris and to lay waste to Germany itself, his orders were not carried out? Surely someone essential to the success of the deflection mission will recognize the danger. Even assurances that the project is designed to destroy some vile enemy nation would probably be disbelieved, because the effects of collision are planetwide (and anyway it's very hard to

Galileo spacecraft approach sequence to the Main Belt asteroid Ida and its moon. The sequence begins at top left, curves around the bottom of the picture, moves to top right and then over to the middle of the page, where only a sliver of surface is imaged. An asteroid approaching impact with the Earth would on successive days show an approach sequence like that at left. Note the first image is barely more than a point. The satellite, the first to be detected for an asteroid, is apparent by about the tenth image. Courtesy NASA and Alfred McEwan, USGS.

make sure your asteroid excavates its monster crater in a particularly deserving nation).

But now imagine a totalitarian state not overrun by enemy troops, but one thriving and self-confident. Imagine a tradition in which orders are obeyed without question. Imagine that those involved in the operation are supplied a cover story: The asteroid is about to impact the Earth, and it is their job to deflect it—but in order not to worry people needlessly, the operation must be performed in secret. In a military setting with a command hierarchy firmly in place, compartmentalization of knowledge, general secrecy, and a cover story, can we be confident that even apocalyptic orders would be disobeyed? Are we really sure that in the next decades and centuries and millennia, nothing like this might happen? How sure are we?

It's no use saying that all technologies can be used for good or for ill. That is certainly true, but when the "ill" achieves a sufficiently apocalyptic scale, we may have to set limits on which technologies may be developed. (In a way we do this all the time, because we can't afford to develop all technologies. Some are favored and some are not.) Or constraints may have to be levied by the community of nations on madmen and autarchs and fanaticism.

Tracking asteroids and comets is prudent, it's good science, and it doesn't cost much. But, knowing our weaknesses, why would we even consider now developing the technology to deflect small worlds? For safety, shall we imagine this technology in the hands of many nations, each providing checks and balances against misuse by another? This is nothing like the old nuclear balance of terror. It hardly inhibits some madman intent on global catastrophe to know that if he does not hurry, a rival may beat him to it. How confident can we be that the community of nations will be able to detect a cleverly designed, clandestine asteroid deflection in time to do something about it? If such a technology were developed, can any international safeguards be envisioned that have a reliability commensurate with the risk?

Even if we restrict ourselves merely to surveillance, there's a risk. Imagine that in a generation we characterize the orbits of 30,000 objects of 100-meter diameter or more, and that this information is publicized, as of course it should be. Maps will be published showing near-Earth space black with the orbits of asteroids and comets, 30,000 swords of Damocles hanging over our heads—ten times more than the number of stars visible to the naked eye under conditions of optimum atmospheric clarity. Public anxiety might be much greater in such a time of knowledge than in our current age of ignorance. There might be irresistible public pressure to develop means to mitigate even nonexistent threats, which would then feed the danger that deflection technology would be misused. For this reason, asteroid discovery and surveillance may not be a mere neutral tool of future policy, but rather a kind of booby trap. To me, the only foreseeable solution is a combination of accurate orbit estimation, realistic threat assessment, and effective public education—so that in democracies at least, the citizens can make their own, informed decisions. This is a job for NASA.

Near-Earth asteroids, and means of altering their orbits, are being looked at seriously. There is some sign that officials in the Department of Defense and the weapons laboratories are beginning to understand that there may be real dangers in planning to push asteroids around. Civilian and military scientists have met to discuss the subject. On first hearing about the asteroid hazard, many people think of it as a kind of Chicken Little fable; Goosey-Lucy, newly arrived and in great excitement, is communicating the urgent news that the sky is falling. The tendency to dismiss the prospect of any catastrophe that we have not personally witnessed is in the long run very foolish. But in this case it may be an ally of prudence.

MEANWHILE WE MUST STILL FACE the deflection dilemma. If we develop and deploy this technology, it may do us in. If we don't, some asteroid or comet may do us in. The resolution of the dilemma hinges, I think, on the fact that the likely timescales of the two dangers are very different—short for the former, long for the latter.

I like to think that our future involvement with near-Earth asteroids will go something like this: From ground-based observatories, we discover all the big ones, plot and monitor their orbits, determine rotation rates and compositions. Scientists are diligent

Sometime in the twenty-first century: As a small asteroid passes nearby, it is greeted by explorers from Earth. The human species may then be in the course of carefully examining these worldlets and—after it is clear that even quite unlikely circumstances will not result in the technology being misused— beginning to experiment with asteroid deflection. Painting by Don Davis.

in explaining the dangers—neither exaggerating nor muting the prospects. We send robotic spacecraft to fly by a few selected bodies, orbit them, land on them, and return surface samples to laboratories on Earth. Eventually we send humans. (Because of the low gravities, they will be able to make standing broad jumps of ten kilometers or more into the sky, and lob a baseball into orbit around the asteroid.) Fully aware of the dangers, we make no attempts to alter trajectories until the potential for misuse of world-altering technologies is much less. That might take a while.

If we're too quick in developing the technology to move worlds around, we may destroy ourselves; if we're too slow, we will surely destroy ourselves. The reliability of world political organizations and the confidence they inspire will have to make significant strides before they can be trusted to deal with a problem of this seriousness. At the same time, there seems to be no acceptable national solution. Who would feel comfortable with the means of world destruction in the hands of some dedicated (or even potential) enemy nation, whether or not our nation had comparable powers? The existence of interplanetary collision hazards, when widely understood, works to bring our species together. When facing a common danger, we humans have sometimes reached heights widely thought impossible; we have set aside our differences—at least until the danger passed.

But *this* danger never passes. The asteroids, gravitationally churning, are slowly altering their orbits; without warning, new comets come careening toward us from the transplutonian darkness. There will always be a need to deal with them in a way that does not endanger us. By posing two different classes of peril—one natural, the other human-made—the small near-Earth worlds provide a new and potent motivation to create effective transnational institutions and to unify the human species. It's hard to see any satisfactory alternative.

In our usual jittery, two-steps-forward-one-step-back mode, we are moving toward unification anyway. There are powerful influences deriving from transportation and communications technologies, the interdependent world economy, and the global environmental crisis. The impact hazard merely hastens the pace.

Eventually, cautiously, scrupulously careful to attempt nothing with asteroids that could inadvertently cause a catastrophe on Earth, I imagine we will begin to learn how to change the orbits of little nonmetallic worlds, smaller than 100 meters across. We begin with smaller explosions and slowly work our way up. We gain experience in changing the orbits of various asteroids and comets of different compositions and strengths. We try to determine which ones can be pushed around and which cannot. By the twenty-second century, perhaps, we move small worlds around the Solar System, using (see next chapter) not nuclear explosions but nuclear fusion engines or their equivalents. We insert small asteroids made of precious and industrial metals into Earth orbit. Gradually we develop a defensive technology to deflect a large asteroid or comet that might in the foreseeable future hit the Earth, while, with meticulous care, we build layers of safeguards against misuse.

Since the danger of misusing deflection technology seems so much greater than the danger of an imminent impact, we can afford to wait, take precautions, rebuild political institutions—for decades certainly, probably centuries. If we play our cards right and are not unlucky, we can pace what we do up there by what progress we're making down here. The two are in any case deeply connected.

The asteroid hazard forces our hand. Eventually, we must establish a formidable human presence throughout the inner Solar System. On an issue of this importance I do not think we will be content with purely robotic means of mitigation. To do so safely we must make changes in our political and international systems. While much about our future is cloudy, this conclusion seems a little more robust, and independent of the vagaries of human institutions.

In the long term, even if we were not the descendants of professional wanderers, even if we were not inspired by exploratory passions, some of us would still have to leave the Earth—simply to ensure the survival of all of us. And once we're out there, we'll need bases, infrastructures. It would not be very long before some of us were living in artificial habitats and on other worlds. This is the first of two missing arguments, omitted in our discussion of missions to Mars, for a permanent human presence in space.

OTHER PLANETARY SYSTEMS must face their own impact hazards—because small primordial worlds, of which asteroids and comets are remnants, are the stuff out of which planets form there as well. After the planets are made, many of these planetesimals are left over. The average time between civilization-threatening impacts on Earth is perhaps 200,000 years, twenty times the age of our civilization. Very different waiting times may pertain to extraterrestrial civilizations, if they exist, depending on such factors as the physical and chemical characteristics of the planet and its biosphere, the biological and social nature of the civilization, and of course the collision rate itself. Planets with higher atmospheric pressures will be protected against somewhat larger impactors, although the pressure cannot be much greater before greenhouse warming and other consequences make life improbable. If the gravity is much less than on Earth, impactors will make less energetic collisions and the hazard will be reduced—although it cannot be reduced very much before the atmosphere escapes to space.

The impact rate in other planetary systems is uncertain. Our system contains two major populations of small bodies that feed potential impactors into Earth-crossing orbits. Both the existence of the source populations and the mechanisms that maintain the collision rate depend on how worlds are distributed. For example, our Oort Cloud seems to have been populated by gravitational ejections of icy worldlets from the vicinity of Uranus and Neptune. If there are no planets that play the role of Uranus and Neptune in systems otherwise like our own, their Oort Clouds may be much more thinly populated. Stars in open and globular stellar clusters, stars in double or multiple systems, stars closer to the center of the Galaxy, stars experiencing more frequent encounters with Giant Molecular Clouds in interstellar space, may all experience higher impact fluxes at their terrestrial planets. The cometary flux might be hundreds or thousands of times more at the Earth had the planet Jupiter never formed—according to a calculation by George Wetherill of the Carnegie Institution of Washington. In systems without Jupiter-like planets, the gravitational shield against comets is down, and civilization-threatening impacts much more frequent.

To a certain extent, increased fluxes of interplanetary objects might increase the rate of evolution, as the mammals that flourished and diversified after the Cretaceous-Tertiary collision wiped out the dinosaurs. But there must be a point of diminishing returns: Clearly, some flux is too high for the continuance of any civilization.

One consequence of this train of argument is that, even if civilizations commonly arise on planets throughout the Galaxy, few of them will be both long-lived and nontechnological. Since hazards from asteroids and comets must apply to inhabited planets all over the Galaxy, if there are such, intelligent beings everywhere will have to unify their home worlds politically, leave their planets, and move small nearby worlds around. Their eventual choice, as ours, is spaceflight or extinction.

REMAKING THE PLANETS

> Who could deny that man could somehow also make the heavens,
> could he only obtain the instruments and the heavenly material?
> —MARSILIO FICINO, "THE SOUL OF MAN" (CA. 1474)

In the midst of the Second World War, a young American writer named Jack Williamson envisioned a populated Solar System. In the twenty-second century, he imagined, Venus would be settled by China,* Japan, and Indonesia; Mars by Germany; and

OPPOSITE: Mars in the course of terraforming, as seen from its moon Phobos. The Valles Marineris is filled with liquid water. Note the lights of cities in the night hemisphere. Painting © David A. Hardy.

* In the real world, Chinese space officials are proposing to send a two-person astronaut capsule into orbit by the turn of the century. It would be propelled by a modified *Long March 2E* rocket and be launched from the Gobi Desert. If the Chinese economy exhibits even moderate continuing growth—much less the exponential growth that marked it in the early to mid-1990s—China may be one of the world's leading space powers by the middle of the twenty-first century. Or earlier.

Surface or underground habitats on asteroids or the Moon seem technically feasible by the middle to late twenty-first century. Many resources could be supplied from the world itself. Because the gravity is so low, human-powered flight would be easy. Painting by Pat Rawlings; copyright Pat Rawlings 1986.

the moons of Jupiter by Russia. Those who spoke English, the language in which Williamson was writing, were confined to the asteroids—and of course the Earth.

The story, published in *Astounding Science Fiction* in July 1942, was called "Collision Orbit" and written under the pseudonym Will Stewart. Its plot hinged on the imminent collision of an uninhabited asteroid with a colonized one, and the search for a means of altering the trajectories of small worlds. Although no one on Earth was endangered, this may have been the first appearance, apart from newspaper comic strips, of asteroid collisions as a threat to humans. (*Comets* impacting the Earth had been a staple peril.)

The environments of Mars and Venus were poorly understood in the early 1940s; it was conceivable that humans could live there without elaborate life-support systems. But the asteroids were another matter. It was well known, even then, that asteroids were small, dry, airless worlds. If they were to be inhabited, especially by large numbers of people, these little worlds would somehow have to be fixed.

In "Collision Orbit," Williamson portrays a group of "spatial engineers," able to render such barren outposts clement. Coining a word, Williamson called the process of metamorphosis into an Earthlike world "terraforming." He knew that the low gravity on an asteroid means that any atmosphere generated or transported there would quickly escape to space. So his key terraforming technology was "paragravity," an artificial gravity that would hold a dense atmosphere.

As nearly as we can tell today, paragravity is a physical impossibility. But we can imagine domed, transparent habitats on the surfaces of asteroids, as suggested by Konstantin Tsiolkovsky, or communities established in the *insides* of asteroids, as outlined in the 1920s by the British scientist J. D. Bernal. Because asteroids are small and their gravities low, even massive subsurface construction might be comparatively easy. If a tunnel were dug clean through, you could jump in at one end and emerge some 45 minutes later at the other, oscillating up and down along the full diameter of this world indefinitely. Inside the right kind of asteroid, a carbonaceous one, you can find materials for manufacturing stone, metal, and plastic construction and plentiful water—all you might need to build a subsurface closed ecological system, an underground garden. Implementation would require a significant step beyond what we have today, but—unlike "paragravity"—nothing in such a scheme seems impossible. All the elements can be found in contemporary technology. If there were sufficient reason, a fair number of us could be living on (or in) asteroids by the twenty-second century.

They would of course need a source of power, not just to sustain themselves, but, as Bernal suggested, to move their asteroidal homes around. (It does not seem so big a step from explosive alteration of asteroid orbits to a more gentle means of propulsion a century or two later.) If an oxygen atmosphere were generated from chemically bound water, then organics could be burned to generate power, just as fossil fuels are burned on the Earth today. Solar power could be considered, although for the main-belt asteroids the intensity of sunlight is only about 10 percent what it is on Earth. Still, we could imagine vast fields of solar panels covering the surfaces of inhabited asteroids and converting sunlight into electricity. Photovoltaic technology is routinely used in Earth-orbiting spacecraft, and is in increasing use on the surface of the Earth today. But while that might be enough to warm and light the homes of these descendants, it does not seem adequate to change asteroid orbits.

For that, Williamson proposed using anti-matter. Anti-matter is just like ordinary matter, with one significant difference. Consider hydrogen: An ordinary hydrogen atom consists of a positively charged proton on the inside and a negatively charged electron on the outside. An atom of anti-hydrogen consists of a negatively charged proton on the inside and a positively charged electron (also called a positron) on the outside. The protons, whatever the sign of their charges, have the same mass; and the electrons, whatever the sign of *their* charges, have the same mass. Particles with opposite charges attract. A hydrogen atom and an anti-hydrogen atom are both stable, because in both cases the positive and negative electrical charges precisely balance.

Anti-matter is not some hypothetical construct from the perfervid musings of science fiction writers or theoretical physicists. Anti-matter exists. Physicists make it in nuclear accelerators; it can be found in high-energy cosmic rays. So why don't we hear more about it? Why has no one held up a lump of anti-matter for our inspection? Because matter and anti-matter, when brought into contact, violently annihilate each other, disappearing in an intense burst of gamma rays. We cannot tell whether something is made of matter or anti-matter just by looking at it. The spectroscopic properties of, for example, hydrogen and anti-hydrogen are identical.

Albert Einstein's answer to the question of why we see only matter and not anti-matter was, "Matter won"—by which he meant that in our sector of the Universe at least, after almost all

OPPOSITE: One small way that humans have tried to put their stamp on the Solar System is by naming features on other worlds after their culture heroes. In these U.S. Geological Survey shaded relief maps of Mercury we see (above) craters named, by the International Astronomical Union, after Peter Illich Tchaikovski, Antonin Dvořàk, Homer, Henrik Ibsen, Gerard Kuiper, Herman Melville, Henri Matisse, Marcel Proust, and Raphael, as well as valleys named for the Arecibo (Cornell) and Goldstone (Jet Propulsion Laboratory) radar observatories. In the projection at bottom, almost exactly at the South Pole, is the crater Chao Meng-Fu, where radar data suggest water ice may be hiding in permanently shadowed regions.

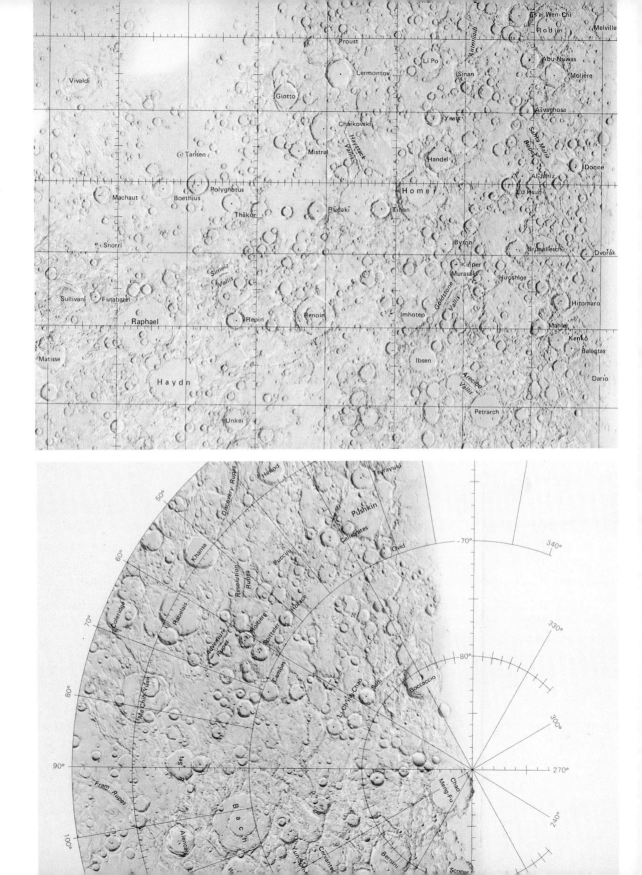

the matter and anti-matter interacted and annihilated each other long ago, there was some of what we call ordinary matter left over.[*] As far as we can tell today, from gamma ray astronomy and other means, the Universe is made almost entirely of matter. The reason for this engages the deepest cosmological issues, which need not detain us here. But if there was only a one-particle-in-a-billion difference in the preponderance of matter over anti-matter at the beginning, even this would be enough to explain the Universe we see today.

Williamson imagined that humans in the twenty-second century would move asteroids around by the controlled mutual annihilation of matter and anti-matter. The resulting gamma rays, if collimated, would make a potent rocket exhaust. The anti-matter would be available in the main asteroid belt (between the orbits of Mars and Jupiter), because this was his explanation for the *existence* of the asteroid belt. In the remote past, he proposed, an intruder anti-matter anti-worldlet arrived in the Solar System from the depths of space, impacted, and annihilated what was then an Earthlike planet, fifth from the Sun. The fragments of this mighty collision are the asteroids, and some of them are still made of anti-matter. Harness an anti-asteroid—Williamson recognized that this might be tricky—and you can move worlds around at will.

At the time, Williamson's ideas were futuristic, but far from foolish. Some of "Collision Orbit" can be considered visionary. Today, however, we have good reason to believe that there are no significant amounts of anti-matter in the Solar System, and that the asteroid belt, far from being a fragmented terrestrial planet, is an enormous array of small bodies prevented (by the gravitational tides of Jupiter) from forming an Earthlike world.

However, we do generate (very) small amounts of anti-matter in nuclear accelerators today, and we will probably be able to manufacture much larger amounts by the twenty-second century. Be-

---

[*] If it had been the other way, then we and everything else in this part of the Universe would be made of anti-matter. We would, of course, call it matter—and the idea of worlds and life made of that other kind of material, the stuff with the electrical charges reversed, we'd consider wildly speculative.

cause it is so efficient—converting *all* of the matter into energy, $E = mc^2$, with 100 percent efficiency—perhaps anti-matter engines will be a practical technology by then, vindicating Williamson. Failing that, what energy sources can we realistically expect to be available to reconfigure asteroids, to light them, warm them, and move them around?

The Sun shines by jamming protons together and turning them into helium nuclei. Energy is released in the process, although with less than 1 percent the efficiency of the annihilation of matter and anti-matter. But even proton-proton reactions are far beyond anything we can realistically imagine for ourselves in the near future. The required temperatures are much too high. Instead of jamming protons together, though, we might use heavier kinds of hydrogen. We already do so in thermonuclear weapons. Deuterium is a proton bound by nuclear forces to a neutron; tritium is a proton bound by nuclear forces to two neutrons. It seems likely that in another century we will have practical power schemes that involve the controlled fusion of deuterium and tritium, and of deuterium and helium. Deuterium and tritium are present as minor constituents in water (on Earth and other worlds). The kind of helium needed for fusion, ^{3}He (two protons and a neutron make up its nucleus), has been implanted over billions of years by the solar wind in the surfaces of the asteroids. These processes are not nearly as efficient as the proton-proton reactions in the Sun, but they could provide enough power to run a small city for a year from a lode of ice only a few meters in size.

Fusion reactors seem to be coming along too slowly to play a major role in solving, or even significantly mitigating, global warming. But by the twenty-second century, they ought to be widely available. With fusion rocket engines, it will be possible to move asteroids and comets around the inner Solar System, taking a main-belt asteroid, for example, and inserting it into orbit around the Earth. A world 10 kilometers across could be transported from Saturn, say, to Mars through nuclear burning of the hydrogen in an icy comet a kilometer across. (Again, I'm assuming a time of much greater political stability and safety.)

PUT ASIDE FOR THE MOMENT any qualms you might have about the ethics of rearranging worlds, or our ability to do so without catastrophic consequences. Digging out the insides of worldlets, reconfiguring them for human habitation, and moving them from one place in the Solar System to another seem to be within our grasp in another century or two. Perhaps by then we will have adequate international safeguards as well. But what about transforming the environments not of asteroids or comets, but of planets? Could we live on Mars?

If we wanted to set up housekeeping on Mars, it's easy to see that, in principle at least, we could do it: There's abundant sunlight. There's plentiful water in the rocks and in underground and polar ice. The atmosphere is mostly carbon dioxide. There's a great deal of organic matter on nearby Phobos, which could be plowed out and delivered to Mars below. (Actually, the surface of Phobos is already grooved, as if someone has been there before us—but planetary geologists think they understand how tidal forces or impact cratering might generate such grooves.) It seems likely that in self-contained habitats—perhaps domed enclosures—we could grow crops, manufacture oxygen from water, recycle wastes.

At first we'd be dependent on commodities resupplied from Earth, but in time we'd manufacture more and more of them ourselves. We'd become increasingly self-sufficient. The domed enclosures, even if made of ordinary glass, would let in the visible sunlight and screen out the Sun's ultraviolet rays. With oxygen masks and protective garments—but nothing as bulky and cumbersome as a spacesuit—we could leave these enclosures to go exploring, or to build another domed village and farms.

It seems very evocative of the American pioneering experience, but with at least one major difference: In the early stages, large subsidies are essential. The technology required is too expensive for some poor family, like my grandparents a century ago, to pay their own passage to Mars. The early Martian pioneers will be sent by governments and will have highly specialized skills. But in a generation or two, when children and grandchildren are born there—and especially when self-sufficiency is within reach—that will begin to change. Youngsters born on Mars will be given specialized training in the technology essential for survival in this new environment. The settlers will become less heroic and less exceptional. The full range of human strengths and weaknesses will begin to assert themselves. Gradually, in part because of the difficulty of getting from Earth to Mars, a unique Martian culture will begin to emerge—distinct aspirations and fears tied to the environment they live in, distinct technologies, distinct social problems, distinct solutions—and, as has occurred in every similar circumstance throughout human history, a gradual sense of cultural and political estrangement from the mother world.

Great ships will arrive carrying essential technology from Earth, new families of settlers, scarce resources. It is hard to know, on the basis of our limited knowledge of Mars, whether they will go home empty—or whether they will carry with them something found only on Mars, something considered very valuable on Earth. Initially much of the scientific investigation of samples of the Martian surface will be done on Earth. But in time the scientific study of Mars (and its moons Phobos and Deimos) will be done from Mars.

Eventually—as has happened with virtually every other form of human transportation—interplanetary travel will become ac-

Part of an international effort to prepare Mars for humans: An American astronaut receives a message from home. NASA artwork by Pat Rawlings/SAIC.

Shift change: Crew members of a base on Mars ascend to Martian orbit, where they will board an interplanetary transport for leave back on Earth. NASA artwork by Pat Rawlings/SAIC.

cessible to people of ordinary means: to scientists pursuing their own research projects, to settlers fed up with Earth, even to venturesome tourists. And of course there will be explorers.

If the time ever came when it was possible to make the Martian environment much more Earthlike—so protective garments, oxygen masks, and domed farmlands and cities could be dispensed with—the attraction and accessibility of Mars would be increased many-fold. The same, of course, would be true for any other world that could be engineered so that humans could live there without elaborate contrivances to keep the planetary environment out. We would feel much more comfortable in our adopted home if an intact dome or spacesuit weren't all that stood between us and death. (But perhaps I exaggerate the worries. People who live in the Netherlands seem at least as well adjusted and carefree as other inhabitants of Northern Europe; yet their dikes are all that stand between them and the sea.)

Recognizing the speculative nature of the question and the

limitations in our knowledge, is it nevertheless possible to envision terraforming the planets?

We need look no further than our own world to see that humans are now able to alter planetary environments in a profound way. Depletion of the ozone layer, global warming from an increased greenhouse effect, and global cooling from nuclear war are all ways in which present technology can significantly alter the environment of our world—and in each case as an inadvertent consequence of doing something else. If we had *intended* to alter our planetary environment, we would be fully able to generate still greater change. As our technology becomes more powerful, we will be able to work still more profound changes.

But just as (in parallel parking) it's easier to get out of a parking place than into one, it's easier to destroy a planetary environment than to move it into a narrowly prescribed range of temperatures, pressures, compositions, and so on. We already know

Early stages in the human habitation of Mars, as conceived by Chesley Bonestell. Painting from the author's collection.

of a multitude of desolate and uninhabitable worlds, and—with very narrow margins—only one green and clement one. This is a major conclusion from early in the era of spacecraft exploration of the Solar System. In altering the Earth, or any world with an atmosphere, we must be very careful about positive feedbacks, where we nudge an environment a little bit and it takes off on its own—a little cooling leading to runaway glaciation, as may have happened on Mars, or a little warming to a runaway greenhouse effect, as happened on Venus. It is not at all clear that our knowledge is sufficient to this purpose.

As far as I know, the first suggestion in the scientific literature about terraforming the planets was made in a 1961 article I wrote about Venus. I was pretty sure then that Venus had a surface temperature well above the normal boiling point of water, produced by a carbon dioxide/water vapor greenhouse effect. I imagined seeding its high clouds with genetically engineered microorganisms that would take CO_2, N_2, and H_2O out of the atmosphere and convert them into organic molecules. The more CO_2 removed, the smaller the greenhouse effect and the cooler the surface. The microbes would be carried down through the atmosphere toward the ground, where they would be fried, so water vapor would be returned to the atmosphere; but the carbon from the CO_2 would be converted irreversibly by the high temperatures into graphite or some other involatile form of carbon. Eventually, the temperatures would fall below the boiling point and the surface of Venus would become habitable, dotted with pools and lakes of warm water.

The idea was soon taken up by a number of science fiction authors in the continuing dance between science and science fiction—in which the science stimulates the fiction, and the fiction stimulates a new generation of scientists, a process benefiting both genres. But as the next step in the dance, it is now clear that seeding Venus with special photosynthetic microorganisms will not work. Since 1961 we've discovered that the clouds of Venus are a concentrated solution of sulfuric acid, which makes the genetic engineering rather more challenging. But that in itself is not a fatal flaw. (There are microorganisms that live out their lives in concentrated solutions of sulfuric acid.) Here's the fatal flaw: In 1961

The surface of Venus as imagined by the pioneering space artist Chesley Bonestell in the 1950s. I reproduced this painting in a 1961 scientific paper on Venus, published in the journal *Science,* in which I suggested a way to terraform Venus. Painting from the author's collection.

I thought the atmospheric pressure at the surface of Venus was a few "bars," a few times the surface pressure on Earth. We now know it to be 90 bars, so that if the scheme worked, the result would be a surface buried in hundreds of meters of fine graphite, and an atmosphere made of 65 bars of almost pure molecular oxygen. Whether we would first implode under the atmospheric pressure or spontaneously burst into flames in all that oxygen is an open question. However, long before so much oxygen could build up, the graphite would spontaneously burn back into CO_2, short-circuiting the process. At best, such a scheme can carry the terraforming of Venus only partway.

Let's assume that by the early twenty-second century we have comparatively inexpensive heavy-lift vehicles, so we can carry large payloads to other worlds; abundant and powerful fusion reactors; and well-developed genetic engineering. All three assumptions are likely, given current trends. Could we terraform the planets?[*] James Pollack of NASA's Ames Research Center and I surveyed this problem. Here's a summary of what we found:

VENUS: Clearly the problem with Venus is its massive greenhouse effect. If we could reduce the greenhouse effect almost to zero, the climate might be balmy. But a 90-bar CO_2 atmosphere is oppressively thick. Over every postage stamp–sized square inch of surface, the air weighs as much as six professional football players, piled one on top of another. Making all that go away will take some doing.

Imagine bombarding Venus with asteroids and comets. Each impact would blow away some of the atmosphere. To blow away almost all of it, though, would require using up more big asteroids and comets than there are—at least in the planetary part of the Solar System. Even if that many potential impactors existed, even if we could make them all collide with Venus (this is the overkill approach to the impact hazard problem), think what we would have lost. Who knows what wonders, what practical knowledge they contain? We would also obliterate much of Venus' gorgeous surface geology—which we've just begun to understand, and which may teach us much about the Earth. This is an example of brute-force terraforming. I suggest we want to steer entirely clear of such methods, even if someday we'll be able to afford them (which I very much doubt). We want something more elegant, more subtle, more respectful of the environments of other worlds. A microbial approach has some of those virtues, but does not do the trick, as we've just seen.

We can imagine pulverizing a dark asteroid and spreading the powder through the upper atmosphere of Venus, or carrying such dust up from the surface. This would be the physical equivalent of nuclear winter or the Cretaceous-Tertiary postimpact climate. If the sunlight reaching the ground is sufficiently attenuated, the sur-

---

[*] Williamson, Professor Emeritus of English at Eastern New Mexico University, at age 85 wrote to me that he was "amazed to see how far actual science has come" since he first suggested terraforming. We are accumulating the technology that will one day permit terraforming, but at present all we have are suggestions by and large less groundbreaking than Williamson's original ideas.

face temperature must fall. But by its very nature, this option plunges Venus into deep gloom, with daytime light levels perhaps only as bright as on a moonlit night on Earth. The oppressive, crushing 90-bar atmosphere would remain untouched. Since the emplaced dust would sediment out every few years, the layer would have to be replenished in the same period of time. Perhaps such an approach would be acceptable for short exploratory missions, but the environment generated seems very stark for a self-sustaining human community on Venus.

We could use a giant artificial sunshade in orbit around Venus to cool the surface; but it would be enormously expensive, as well as having many of the deficiencies of the dust layer. However, if the temperatures could be lowered sufficiently, the CO_2 in the atmosphere would rain out. There would be a transitional time of CO_2 oceans on Venus. If those oceans could be covered over to prevent re-evaporation—for example, with water oceans made by melting a large, icy moon transported from the outer Solar System—then the CO_2 might conceivably be sequestered away, and Venus converted into a water (or low-fizz seltzer) planet. Ways have also been suggested to convert the CO_2 into carbonate rock.

Thus all proposals for terraforming Venus are still brute-force, inelegant, and absurdly expensive. The desired planetary metamorphosis may be beyond our reach for a very long time, even if we thought it was desirable and responsible. The Asian colonization of Venus that Jack Williamson imagined may have to be redirected somewhere else.

MARS: For Mars we have just the opposite problem. There's not *enough* greenhouse effect. The planet is a frozen desert. But the fact that Mars seems to have had abundant rivers, lakes, and perhaps even oceans 4 billion years ago—at a time when the Sun was less bright than it is today—makes you wonder if there's some natural instability in the Martian climate, something on hair trigger that once released would all by itself return the planet to its ancient clement state. (Let's note from the start that doing so would destroy Martian landforms that hold key data on the past—especially the laminated polar terrain.)

As we know very well from Earth and Venus, carbon dioxide

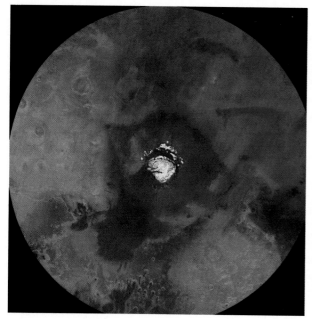
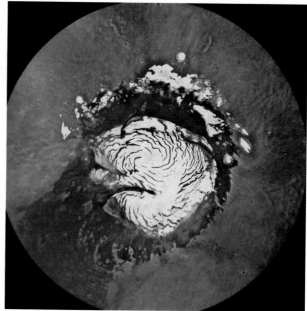

ABOVE LEFT: Looking down on the north polar cap of Mars. *Viking* photomosaic, courtesy USGS/NASA.

ABOVE RIGHT: Close-up of the north polar cap. The amount of carbon dioxide locked away in the polar caps of Mars seems insufficient to account for the dense atmosphere deduced for ancient Mars. Perhaps there are abundant carbonates in the Martian soil. However, between carbon dioxide in the polar caps and the Martian soil and other gases that can be manufactured on Mars, it now seems possible that a sufficient greenhouse effect could be generated to transform the Martian environment toward much more Earthlike conditions. *Viking* photomosaic, courtesy USGS/NASA.

is a greenhouse gas. There are carbonate minerals found on Mars, and dry ice in one of the polar caps. They could be converted into CO_2 gas. But to make enough of a greenhouse effect to generate comfortable temperatures on Mars would require the entire surface of the planet to be plowed up and processed to a depth of kilometers. Apart from the daunting obstacles in practical engineering that this represents—fusion power or no fusion power—and the inconvenience to whatever self-contained, closed ecological systems humans had already established on the planet, it would also constitute the irresponsible destruction of a unique scientific resource and database, the Martian surface.

What about other greenhouse gases? We might take chlorofluorocarbons (CFCs or HCFCs) to Mars after manufacturing them on Earth. These artificial substances, so far as we know, are found nowhere else in the Solar System. We can certainly imagine manufacturing enough CFCs on Earth to warm Mars, because by *accident* in a few decades with present technology on Earth we've managed to synthesize enough to contribute to global warming on our planet. Transportation to Mars would be expensive, though: Even using *Saturn V*- or *Energiya*-class boosters, it would require at least a launch a day for a century. But perhaps they could be manufactured from fluorine-containing minerals on Mars.

There is, in addition, a serious drawback: On Mars as on

Earth, abundant CFCs would prevent formation of an ozone layer. CFCs might bring Martian temperatures into a clement range, but guarantee that the solar ultraviolet hazard would remain extremely serious. Perhaps the solar ultraviolet light could be absorbed by an atmospheric layer of pulverized asteroidal or surface debris injected in carefully titrated amounts above the CFCs. But now we're in the troubling circumstance of having to deal with propagating side effects, each of which requires its own large-scale technological solution.

A third possible greenhouse gas for warming Mars is ammonia (NH_3). Only a little ammonia would be enough to warm the Martian surface to above the freezing point of water. In principle, this might be done by specially engineered microorganisms that would convert Martian atmospheric N_2 to NH_3 as some microbes do on Earth, but do it under Martian conditions. Or the same conversion might be done in special factories. Alternatively, the nitrogen required could be carried to Mars from elsewhere in the Solar System. (N_2 is the principal constituent in the atmospheres of both Earth and Titan.) Ultraviolet light would convert ammonia back into N_2 in about 30 years, so there would have to be a continuous resupply of NH_3.

The terraces of the Martian north polar cap. Dust and ice are interlayered, potentially holding vital information on the history of past climate change on Mars. Painting by Ron Miller.

346 PALE BLUE DOT

A judicious combination of CO_2, CFC, and NH_3 greenhouse effects on Mars looks as if it might be able to bring surface temperatures close enough to the freezing point of water for the second phase of Martian terraforming to begin—temperatures rising further due to substantial water vapor in the air, widespread production of O_2 by genetically engineered plants, and fine-tuning the surface environment. Microbes and larger plants and animals could be established on Mars before the overall environment was suitable for unprotected human settlers.

Terraforming Mars is plainly much easier than terraforming Venus. But it is still very expensive by present standards, and environmentally destructive. If there were sufficient justification, though, perhaps the terraforming of Mars could be under way by the twenty-second century.

THE MOONS OF JUPITER AND SATURN: Terraforming the satellites of the Jovian planets presents varying degrees of difficulty. Perhaps the easiest to contemplate is Titan. It already has an atmosphere, made mainly of N_2 like the Earth's, and is much closer to terrestrial atmospheric pressures than either Venus or Mars. Moreover, important greenhouse gases, such as NH_3 and H_2O, are almost certainly frozen out on its surface. Manufacture of initial greenhouse gases that do not freeze out at present Titan temperatures plus direct warming of the surface by nuclear fusion could, it seems, be the key early steps to one day terraform Titan.

IF THERE WERE A COMPELLING REASON for terraforming other worlds, this greatest of engineering projects might be feasible on the timescale we've been describing—certainly for asteroids, possibly for Mars, Titan, and other moons of the outer planets, and probably not for Venus. Pollack and I recognized that there are those who feel a powerful attraction to the idea of rendering other worlds in the Solar System suitable for human habitation—in establishing observatories, exploratory bases, communities, and homesteads there. Because of its pioneering history, this may be a particularly natural and attractive idea in the United States.

In any case, massive alteration of the environments of other worlds can be done competently and responsibly only when we have a much better understanding of those worlds than is available today. Advocates of terraforming must first become advocates of the long-term and thorough scientific exploration of other worlds.

Perhaps when we really understand the difficulties of terraforming, the costs or the environmental penalties will prove too steep, and we will lower our sights to domed or subsurface cities or other local, closed ecological systems, greatly improved versions of Biosphere II, on other worlds. Perhaps we will abandon the dream of converting the surfaces of other worlds to something approaching the Earth's. Or perhaps there are much more elegant, cost-effective, and environmentally responsible ways of terraforming that we have not yet imagined.

But if we are seriously to pursue the matter, certain questions ought to be asked: Given that any terraforming scheme entails a balance of benefits against costs, how certain must we be that key scientific information will not thereby be destroyed before proceeding? How much understanding of the world in question do we need before planetary engineering can be relied upon to produce the desired end state? Can we guarantee a long-term human commitment to maintain and replenish an engineered world, when human political institutions are so short-lived? If a world is even conceivably inhabited—perhaps only by microorganisms—do humans have a right to alter it? What is our responsibility to

Artist's conception of the terraforming in the far future of a moon of a Jovian planet. Painting by Michael Carroll.

preserve the worlds of the Solar System in their present wilderness states for future generations—who may contemplate uses that today we are too ignorant to foresee? These questions may perhaps be encapsulated into a final question: Can we, who have made such a mess of *this* world, be trusted with others?

It is just conceivable that some of the techniques that might eventually terraform other worlds might be applied to ameliorate the damage we have done to this one. Considering the relative urgencies, a useful indication of when the human species is ready to consider terraforming seriously is when we have put our own world right. We can consider it a test of the depth of our understanding and our commitment. The first step in engineering the Solar System is to guarantee the habitability of the Earth.

Then we'll be ready to spread out to asteroids, comets, Mars, the moons of the outer Solar System, and beyond. Jack Williamson's prediction that this will begin to come about by the twenty-second century may not be far off the mark.

THE NOTION OF OUR DESCENDANTS living and working on other worlds, and even moving some of them around for their convenience, seems the most extravagant science fiction. Be realistic, a voice inside my head counsels. But this *is* realistic. We're on the cusp of the technology, near the midpoint between impossible and routine. It's easy to be conflicted about it. If we don't do something awful to ourselves in the interim, in another century terraforming may seem no more impossible than a human-tended space station does today.

I think the experience of living on other worlds is bound to change us. Our descendants, born and raised elsewhere, will naturally begin to owe primary loyalty to the worlds of their birth, whatever affection they retain for the Earth. Their physical needs, their methods of supplying those needs, their technologies, and their social structures will all have to be different.

A blade of grass is a commonplace on Earth; it would be a miracle on Mars. Our descendants on Mars will know the value of a patch of green. And if a blade of grass is priceless, what is the value of a human being? The American revolutionary Tom Paine, in describing his contemporaries, had thoughts along these lines:

The Association of Space Explorers is one of the most exclusive organizations on Earth: You must have traveled in space to be a member. This poster, signed by astronauts and cosmonauts from 25 nations, was prepared for the 1992 annual meeting of the Association, which was devoted to human missions to Mars. As is true for Mars itself, in this poster which way is up is merely a matter of convention. From the author's collection.

The wants which necessarily accompany the cultivation of a wilderness produced among them a state of society which countries long harassed by the quarrels and intrigues of governments had neglected to cherish. In such a situation man becomes what he ought to be. He sees his species . . . as kindred.

Having seen at first hand a procession of barren and desolate worlds, it will be natural for our spacefaring descendants to cherish life. Having learned from the tenure of our species on Earth, they may wish to apply those lessons to other worlds—to spare generations to come the avoidable suffering that their ancestors were obliged to endure, and to draw upon our experience and our mistakes as we begin our open-ended evolution into space.

CHAPTER 20

DARKNESS

Far away, hidden from the eyes of daylight, there are watchers in the skies.

—EURIPIDES, *THE BACCHAE* (CA. 406 B.C.)

A s children, we fear the dark. Anything might be out there. The unknown troubles us.

Ironically, it is our fate to *live* in the dark. This unexpected finding of science is only about three centuries old. Head out from the Earth in any direction you choose, and—after an initial flash of blue and a longer wait while the Sun fades—you are surrounded by blackness, punctuated only here and there by the faint and distant stars.

Even after we are grown, the darkness retains its power to frighten us. And so there are those who say we should not inquire too closely into who else might be living in that darkness. Better not to know, they say.

There are 400 billion stars in the Milky Way Galaxy. Of this immense multitude, could it be that our humdrum Sun is the only

OPPOSITE: Radio telescopes examine the night sky, which looks very different in radio waves than in ordinary visible light. Many of the "stars" are not stars at all, but bright radio galaxies and quasars billions of light-years away. Amid such radio sources and the radio noise of our own technical civilization, can we find in the darkness evidence of other civilizations in space? Image courtesy National Radio Astronomy Observatory.

one with an inhabited planet? Maybe. Maybe the origin of life or intelligence is exceedingly improbable. Or maybe civilizations arise all the time, but wipe themselves out as soon as they are able.

Or, here and there, peppered across space, orbiting other suns, maybe there are worlds something like our own, on which other beings gaze up and wonder as we do about who else lives in the dark. Could the Milky Way be rippling with life and intelligence—worlds calling out to worlds—while we on Earth are alive at the critical moment when we first decide to listen?

Our species has discovered a way to communicate through the dark, to transcend immense distances. No means of communication is faster or cheaper or reaches out farther. It's called radio.

After billions of years of biological evolution—on their planet and ours—an alien civilization cannot be in technological lockstep with us. There have been humans for more than twenty thousand centuries, but we've had radio only for about one century. If alien civilizations are behind us, they're likely to be too far behind to have radio. And if they're ahead of us, they're likely to be far ahead of us. Think of the technical advances on our world over just the last few centuries. What is for us technologically difficult or impossible, what might seem to us like magic, might for them be trivially easy. They might use other, very advanced means to communicate with their peers, but they would know about radio as an approach to newly emerging civilizations. Even with no more than our level of technology at the transmitting and receiving ends, we could communicate today across much of the Galaxy. They should be able to do much better.

If they exist.

But our fear of the dark rebels. The idea of alien beings troubles us. We conjure up objections:

"It's too expensive." But, in its fullest modern technological expression, it costs less than one attack helicopter a year.

"We'll never understand what they're saying." But, because the message is transmitted by radio, we and they must have radio physics, radio astronomy, and radio technology in common. The laws of Nature are the same everywhere; so science itself provides

a means and language of communication even between very different kinds of beings—provided they both have science. Figuring out the message, if we're fortunate enough to receive one, may be much easier than acquiring it.

"It would be demoralizing to learn that our science is primitive." But by the standards of the next few centuries, at least some of our present science will be considered primitive, extraterrestrials or no extraterrestrials. (So will some of our present politics, ethics, economics, and religion.) To go beyond present science is one of the chief goals of science. Serious students are not commonly plunged into fits of despair on turning the pages of a textbook and discovering that some further topic is known to the author but not yet to the student. Usually the students struggle a little, acquire the new knowledge, and, following an ancient human tradition, continue to turn the pages.

"All through history advanced civilizations have ruined civilizations just slightly more backward." Certainly. But malevolent aliens, should they exist, will not discover our existence from the fact that we listen. The search programs only receive; they do not send.*

THE DEBATE IS, for the moment, moot. We are now, on an unprecedented scale, listening for radio signals from possible other civilizations in the depths of space. Alive today is the first generation of scientists to interrogate the darkness. Conceivably it might also be the last generation before contact is made—and this the last moment before we discover that someone in the darkness is calling out to us.

---

* Surprisingly many people, including *New York Times* editorialists, are concerned that once extraterrestrials know where we are, they will come here and eat us. Put aside the profound biological differences that must exist between the hypothetical aliens and ourselves; imagine that we constitute an interstellar gastronomic delicacy. Why transport large numbers of us to alien restaurants? The freightage is enormous. Wouldn't it be better just to steal a few humans, sequence our amino acids or whatever else is the source of our delectability, and then just synthesize the identical food product from scratch?

This quest is called the Search for Extraterrestrial Intelligence (SETI). Let me describe how far we've come.

The first SETI program was carried out by Frank Drake at the National Radio Astronomy Observatory in Greenbank, West Virginia, in 1960. He listened to two nearby Sun-like stars for two weeks at one particular frequency. ("Nearby" is a relative term: The nearest was 12 light-years—70 trillion miles—away.)

Almost at the moment Drake pointed the radio telescope and turned the system on, he picked up a very strong signal. Was it a message from alien beings? Then it went away. If the signal disappears, you can't scrutinize it. You can't see if, because of the Earth's rotation, it moves with the sky. If it's not repeatable, you've learned almost nothing from it—it might be terrestrial radio interference, or a failure of your amplifier or detector . . . or an alien signal. Unrepeatable data, no matter how illustrious the scientist reporting them, are not worth much.

Weeks later, the signal was detected again. It turned out to be a military aircraft broadcasting on an unauthorized frequency. Drake reported negative results. But in science a negative result is not at all the same thing as a failure. His great achievement was to show that modern technology is fully able to listen for signals from hypothetical civilizations on the planets of other stars.

Since then there've been a number of attempts, often on time borrowed from other radio telescope observing programs, and almost never for longer than a few months. There've been some more false alarms, at Ohio State, in Arecibo, Puerto Rico, in France, Russia, and elsewhere, but nothing that could pass muster with the world scientific community.

Meanwhile, the technology for detection has been getting cheaper; the sensitivity keeps improving; the scientific respectability of SETI has continued to grow; and even NASA and Congress have become a little less afraid to support it. Diverse, complementary search strategies are possible and necessary. It was clear years ago that if the trend continued, the technology for a comprehensive SETI effort would eventually fall within the reach even of private organizations (or wealthy individuals); and sooner or later, the government would be willing to support a major program.

After 30 years of work, for some of us it's been later rather than sooner. But at last the time has come.

THE PLANETARY SOCIETY—a nonprofit membership organization that Bruce Murray, then the Director of JPL, and I founded in 1980— is devoted to planetary exploration and the search for extraterrestrial life. Paul Horowitz, a physicist at Harvard University, had made a number of important innovations for SETI and was eager to try them out. If we could find the money to get him started, we thought we could continue to support the program by donations from our members.

In 1983 Ann Druyan and I suggested to the filmmaker Steven Spielberg that this was an ideal project for him to support. Breaking with Hollywood tradition, he had in two wildly successful movies conveyed the idea that extraterrestrial beings might not be hostile and dangerous. Spielberg agreed. With his initial support through The Planetary Society, Project META began.

META is an acronym for "Megachannel ExtraTerrestrial Assay." The single frequency of Drake's first system grew to 8.4 million. But each channel, each "station," we tune to has an exceptionally narrow frequency range. There are no known processes out among the stars and galaxies that can generate such sharp radio "lines." If we pick up anything falling into so narrow a channel, it must, we think, be a token of intelligence and technology.

What's more, the Earth turns—which means that any distant radio source will have a sizable apparent motion, like the rising and setting of the stars. Just as the steady tone of a car's horn dips as it drives by, so any authentic extraterrestrial radio source will exhibit a steady drift in frequency due to the Earth's rotation. In contrast, any source of radio interference at the Earth's surface will be rotating at the same speed as the META receiver. META's listening frequencies are continuously changed to compensate for the Earth's rotation, so that any narrow-band signals from the sky will always appear in a single channel. But any radio interference down here on Earth will give itself away by racing through adjacent channels.

The META radio telescope at Harvard, Massachusetts, is 26

meters (84 feet) in diameter. Each day, as the Earth rotates the tele-scope beneath the sky, a swath of stars narrower than the full moon is swept out and examined. Next day, it's an adjacent swath. Over a year, all of the northern sky and part of the southern is observed. An identical system, also sponsored by The Planetary Society, is in operation just outside Buenos Aires, Argentina, to examine the southern sky. So together the two META systems have been ex-ploring the entire sky.

The radio telescope, gravitationally glued to the spinning Earth, looks at any given star for about two minutes. Then it's on to the next. 8.4 million channels sounds like a lot, but remember, each channel is very narrow. All of them together constitute only a few parts in 100,000 of the available radio spectrum. So we have to park our 8.4 million channels somewhere in the radio spectrum for each year of observation, near some frequency that an alien civilization, knowing nothing about us, might nevertheless con-clude we're listening to.

Hydrogen is by far the most abundant kind of atom in the Universe. It's distributed in clouds and as diffuse gas throughout interstellar space. When it acquires energy, it releases some of it by giving off radio waves at a precise frequency of 1420.405751768 megahertz. (One hertz means the crest and trough of a wave arriv-ing at your detection instrument each second. So 1420 megahertz means 1.420 *billion* waves entering your detector every second. Since the wavelength of light is just the speed of light divided by the frequency of the wave, 1420 megahertz corresponds to a wavelength of 21 centimeters.) Radio astronomers anywhere in the Galaxy will be studying the Universe at 1420 megahertz and can anticipate that other radio astronomers, no matter how differ-ent they may look, will do the same.

It's as if someone told you that there's only one station on your home radio set's frequency band, but that no one knows its frequency. Oh yes, one other thing: Your set's frequency dial, with its thin marker you adjust by turning a knob, happens to reach from the Earth to the Moon. To search systematically through this vast radio spectrum, patiently turning the knob, is going to be very time-consuming. Your problem is to set the dial correctly from the

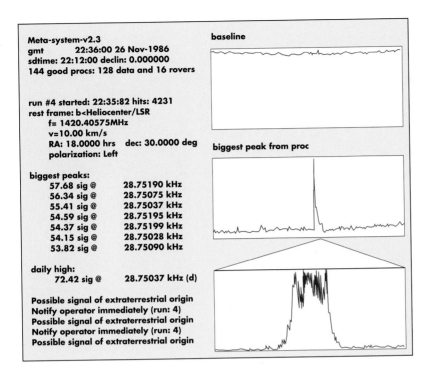

Data and system printout of a candidate signal, Project META, November 26, 1986. From Paul Horowitz and Carl Sagan, *The Astrophysical Journal*, September 20, 1993.

beginning, to choose the right frequency. If you can correctly guess what frequencies that extraterrestrials are broadcasting to us on—the "magic" frequencies—then you can save yourself much time and trouble. These are the sorts of reasons that we first listened, as Drake did, at frequencies near 1420 megahertz, the hydrogen "magic" frequency.

Horowitz and I have published detailed results from five years of full-time searching with Project META and two years of follow-up. We can't report that we found a signal from alien beings. But we did find something puzzling, something that for me in quiet moments, every now and then, raises goose bumps:

Of course, there's a background level of radio noise from Earth—radio and television stations, aircraft, portable telephones, nearby and more distant spacecraft. Also, as with all radio receivers, the longer you wait, the more likely it is that there'll be some random fluctuation in the electronics so strong that it generates a spurious signal. So we ignore anything that isn't *much* louder than the background.

Distribution of META radio sources over the sky. Declination and right ascension are the latitude and longitude sky coordinates used in astronomy. At top are all the candidate signals from five years of observation at the "magic" frequency of 2840 megahertz (the overtone of the 1420-megahertz hydrogen line) that also satisfied previously set criteria for extraterrestrial intelligence, including appearing in a single narrow-band channel (ch). Closer examination permits us to discard most of the data as due to noise in the electronics, or radio frequency interference from the Earth and spacecraft. What remains (bottom) at this frequency are four events. They lie very close to the plane of the Milky Way Galaxy (shown with a dashed line). Two of them lie near the galactic center, marked by an "X." From Paul Horowitz and Carl Sagan, *The Astrophysical Journal*, September 20, 1993.

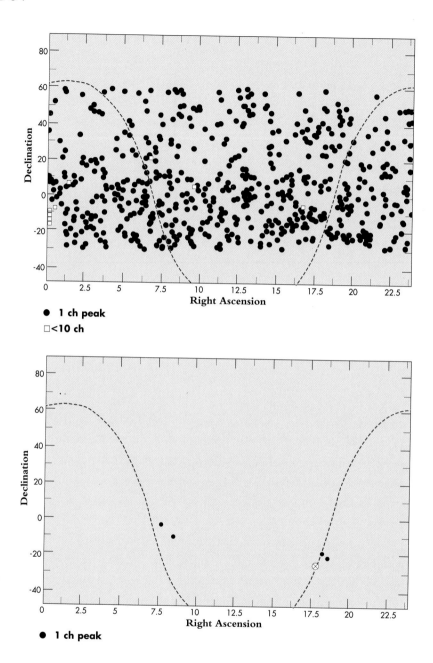

Any strong narrow-band signal that remains in a single channel we take very seriously. As it logs in the data, META automatically tells the human operators to pay attention to certain signals.
Over five years we made some 60 trillion observations at various

frequencies, while examining the entire accessible sky. A few dozen signals survive the culling. These are subjected to further scrutiny, and almost all of them are rejected—for example, because an error has been found by fault-detection microprocessors that examine the signal-detection microprocessors.

What's left—the strongest candidate signals after three surveys of the sky—are 11 "events." They satisfy all but one of our criteria for a genuine alien signal. But the one failed criterion is supremely important: Verifiability. We've never been able to find any of them again. We look back at that part of the sky three minutes later and there's nothing there. We look again the following day: nothing. Examine it a year later, or seven years later, and still there's nothing.

It seems unlikely that every signal we get from alien civilizations would turn itself off a couple of minutes after we begin listening, and never repeat. (How would they know we're paying attention?) But, just possibly, this is the effect of twinkling. Stars twinkle because parcels of turbulent air are moving across the line of sight between the star and us. Sometimes these air parcels act as a lens and cause the light rays from a given star to converge a little, making it momentarily brighter. Similarly, astronomical radio sources may also twinkle—owing to clouds of electrically charged (or "ionized") gas in the great near-vacuum between the stars. We observe this routinely with pulsars.

Imagine a radio signal that's a little below the strength that we could otherwise detect on Earth. Occasionally the signal will by chance be temporarily focused, amplified, and brought within the detectability range of our radio telescopes. The interesting thing is that the lifetimes of such brightening, predicted from the physics of the interstellar gas, *are* a few minutes—and the chance of reacquiring the signal is small. We should really be pointing steadily at these coordinates in the sky, watching them for months.

Despite the fact that none of these signals repeats, there's an additional fact about them that, every time I think about it, sends a chill down my spine: 8 of the 11 best candidate signals lie in or near the plane of the Milky Way Galaxy. The five strongest are in the constellations Cassiopeia, Monoceros, Hydra, and two in Sagittarius—in the approximate direction of the center of the Galaxy. The

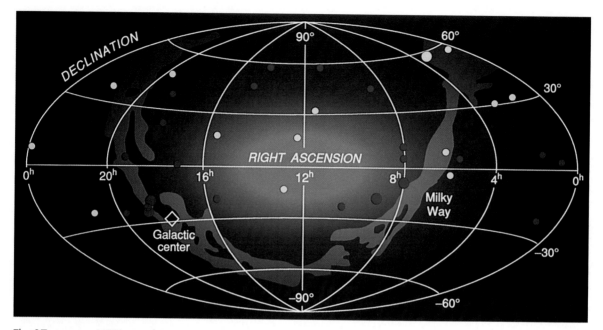

The 37 strongest META signals that survive culling. Yellow dots mark detections at 1420 megahertz and red dots at 2840 megahertz. The large dots mark the five strongest signals. Again, note the concentration of the strongest signals to the plane of the Milky Way. Copyright 1994 by Sky Publishing Corp. Reproduced with permission. Diagram adapted by José R. Díaz, *Sky and Telescope* magazine, from Paul Horowitz and Carl Sagan, *The Astrophysical Journal*, September 20, 1993.

Milky Way is a flat, wheel-like collection of gas and dust and stars. Its flatness is why we see it as a band of diffuse light across the night sky. That's where almost all the stars in our galaxy are. If our candidate signals really were radio interference from Earth or some undetected glitch in the detection electronics, we shouldn't see them preferentially when we're pointing at the Milky Way.

But maybe we had an especially unlucky and misleading run of statistics. The probability that this correlation with the galactic plane is due merely to chance is less than half a percent. Imagine a wall-size map of the sky, ranging from the North Star at the top to the fainter stars toward which the Earth's south pole points at the bottom. Snaking across this wall map are the irregular boundaries of the Milky Way. Now suppose that you were blindfolded and asked to throw five darts at random at the map (with much of the southern sky, inaccessible from Massachusetts, declared off limits). You'd have to throw the set of five darts more than 200 times before, by accident, you got them to fall as closely within the precincts of the Milky Way as the five strongest META signals did. Without repeatable signals, though, there's no way we can conclude that we've actually found extraterrestrial intelligence.

Or maybe the events we've found are caused by some new kind of astrophysical phenomenon, something that nobody has thought of yet, by which not civilizations, but stars or gas clouds (or something) that do lie in the plane of the Milky Way emit strong signals in bafflingly narrow frequency bands.

Let's permit ourselves, though, a moment of extravagant speculation. Let's imagine that *all* our surviving events are in fact due to radio beacons of other civilizations. Then we can estimate—from how little time we've spent watching each piece of sky—how many such transmitters there are in the entire Milky Way. The answer is something approaching a million. If randomly strewn through space, the nearest of them would be a few hundred light years away, too far for them to have picked up our own TV or radar signals yet. They would not know for another few centuries that a technical civilization has emerged on Earth. The Galaxy would be pulsing with life and intelligence, but—unless they're busily exploring huge numbers of obscure star systems—wholly oblivious of what has been happening down here lately. A few centuries from now, after they do hear from us, things might get very interesting. Fortunately, we'd have many generations to prepare.

If, on the other hand, *none* of our candidate signals is an authentic alien radio beacon, then we're forced to the conclusion that very few civilizations are broadcasting, maybe none, at least at our magic frequencies and strongly enough for us to hear:

Consider a civilization like our own, but which dedicated all its available power (about 10 trillion watts) to broadcasting a beacon signal at one of our magic frequencies and to all directions in space. The META results would then imply that there are no such civilizations out to 25 light-years—a volume that encompasses perhaps a dozen Sun-like stars. This is not a very stringent limit. If, in contrast, that civilization were broadcasting directly at our position in space, using an antenna no more advanced than the Arecibo Observatory, then if META has found nothing, it follows that there are no such civilizations anywhere in the Milky Way Galaxy—out of 400 billion stars, not one. But even assuming they would want to, how would they know to transmit in our direction?

Now consider, at the opposite technological extreme, a very

advanced civilization omnidirectionally and extravagantly broadcasting at a power level 10 trillion times greater (10^{26} watts, the entire energy output of a star like the Sun). Then, if the META results are negative, we can conclude not only that there are no such civilizations in the Milky Way, but none out to 70 million light-years—none in M31, the nearest galaxy like our own, none in M33, or the Fornax system, or M81, or the Whirlpool Nebula, or Centaurus A, or the Virgo cluster of galaxies, or the nearest Seyfert galaxies; none among any of the hundred trillion stars in thousands of nearby galaxies. Stake through its heart or not, the geocentric conceit stirs again.

Of course, it might be a token not of intelligence but of stupidity to pour so much energy into interstellar (and intergalactic) communication. Perhaps they have good reasons not to hail all comers. Or perhaps they don't care about civilizations as backward as we are. But still—not one civilization in a hundred trillion stars broadcasting with such power on such a frequency? If the META results are negative, we have set an instructive limit—but whether on the abundance of very advanced civilizations or their communications strategy we have no way of knowing. Even if META has found nothing, a broad middle range remains open—of abundant civilizations, more advanced than we and broadcasting omnidirectionally at magic frequencies. We would not have heard from them yet.

ON OCTOBER 12, 1992—auspiciously or otherwise the 500th anniversary of the "discovery" of America by Christopher Columbus— NASA turned on *its* new SETI program. At a radio telescope in the Mojave Desert, a search was initiated intended to cover the entire sky systematically—like META, making no guesses about which stars are more likely, but greatly expanding the frequency coverage. At the Arecibo Observatory, an even more sensitive NASA study began that concentrated on promising nearby star systems. When fully operational, the NASA searches would have been able to detect much fainter signals than META, and look for kinds of signals that META could not.

The META experience reveals a thicket of background static and radio interference. Quick reobservation and confirmation

of the signal—especially at other, independent radio telescopes—
is the key to being sure. Horowitz and I gave the NASA scientists
the coordinates of our fleeting and enigmatic events. Perhaps they
would be able to confirm and clarify our results. The NASA program was also developing new technology, stimulating ideas, and
exciting schoolchildren. In the eyes of many it was well worth the
$10 million a year being spent on it. But almost exactly a year after
authorizing it, Congress pulled the plug on NASA's SETI program. It cost too much, they said. The post–Cold War U.S. defense
budget is some 30,000 times larger.

The chief argument of the principal opponent of the NASA
SETI program—Senator Richard Bryan of Nevada—was this
[from the *Congressional Record* for September 22, 1993]:

> So far, the NASA SETI Program has found nothing. In
> fact, all the decades of SETI research have found no confirmable
> signs of extraterrestrial life.
>
> Even with the current NASA version of SETI, I do not
> think many of its scientists would be willing to guarantee that
> we are likely to see any tangible results in the [foreseeable] fu-
> ture . . .
>
> Scientific research rarely, if ever, offers guarantees of suc-
> cess—and I understand that—and the full benefits of such re-
> search are often unknown until very late in the process. And I
> accept that, as well.
>
> In the case of SETI, however, the chances of success are so
> remote, and the likely benefits of the program are so limited, that
> there is little justification for 12 million taxpayer dollars to be
> expended for this program.

But how, before we have found extraterrestrial intelligence,
can we "guarantee" that we will find it? How, on the other hand,
can we know that the chances of success are "remote"? And if we
find extraterrestrial intelligence, are the benefits really likely to be
"so limited"? As in all great exploratory ventures, we do not know
what we will find and we don't know the probability of finding it.
If we did, we would not have to look.

SETI is one of those search programs irritating to those who want well-defined cost/benefit ratios. Whether ETI can be found; how long it would take to find it; and what it would cost to do so are all unknown. The benefits might be enormous, but we can't really be sure of that either. It would of course be foolish to spend a major fraction of the national treasure on such ventures, but I wonder if civilizations cannot be calibrated by whether they pay *some* attention to trying to solve the great problems.

Despite these setbacks, a dedicated band of scientists and engineers, centered at the SETI Institute in Palo Alto, California, has decided to go ahead, government or no government. NASA has given them permission to use the equipment already paid for; captains of the electronics industry have donated a few million dollars; at least one appropriate radio telescope is available; and the initial stages of this grandest of all SETI programs is on track. If it can demonstrate that a useful sky survey is possible without being swamped by background noise—and especially if, as is very likely from the META experience, there are unexplained candidate signals—perhaps Congress will change its mind once more and fund the project.

Meanwhile, Paul Horowitz has come up with a new program—different from META, different from what NASA was doing—called BETA. BETA stands for "*Billion*-channel ExtraTerrestrial Assay." It combines narrow-band sensitivity, wide frequency coverage, and a clever way to verify signals as they're detected. If The Planetary Society can find the additional support, this system—much cheaper than the former NASA program—should be on the air soon.

WOULD I LIKE TO BELIEVE that with META we've detected transmissions from other civilizations out there in the dark, sprinkled through the vast Milky Way Galaxy? You bet. After decades of wondering and studying this problem, of course I would. To me, such a discovery would be thrilling. It would change everything. We would be hearing from other beings, independently evolved over billions of years, viewing the Universe perhaps very differently, probably much smarter, certainly not human. How much do they know that we don't?

For me, no signals, no one calling out to us is a depressing prospect. "Complete silence," said Jean-Jacques Rousseau in a different context, "induces melancholy; it is an image of death." But I'm with Henry David Thoreau: "Why should I feel lonely? Is not our planet in the Milky Way?"

The realization that such beings exist and that, as the evolutionary process requires, they must be very different from us, would have a striking implication: Whatever differences divide us down here on Earth are trivial compared to the differences between any of us and any of them. Maybe it's a long shot, but the discovery of extraterrestrial intelligence might play a role in unifying our squabbling and divided planet. It would be the last of the Great Demotions, a rite of passage for our species and a transforming event in the ancient quest to discover our place in the Universe.

In our fascination with SETI, we might be tempted, even without good evidence, to succumb to belief; but this would be self-indulgent and foolish. We must surrender our skepticism only in the face of rock-solid evidence. Science demands a tolerance for ambiguity. Where we are ignorant, we withhold belief. Whatever annoyance the uncertainty engenders serves a higher purpose: It drives us to accumulate better data. This attitude is the difference between science and so much else. Science offers little in the way of cheap thrills. The standards of evidence are strict. But when followed they allow us to see far, illuminating even a great darkness.

TO THE SKY!

The stairs of the sky are let down for him that he may ascend thereon to
heaven. O gods, put your arms under the king: raise him, lift him to the sky.
To the sky! To the sky!

—HYMN FOR A DEAD PHARAOH (EGYPT, CA. 2600 B.C.)

W hen my grandparents were children, the electric light, the automobile, the airplane, and the radio were stupefying technological advances, the wonders of the age. You might hear wild stories about them, but you could not find a single exemplar in that little village in Austria-Hungary, near the banks of the river Bug. But in that same time, around the turn of the last century, there were two men who foresaw other, far more ambitious, inventions—Konstantin Tsiolkovsky, the theoretician, a nearly deaf schoolteacher in the obscure Russian town of

The star field at the Southern Cross. Copyright A. Fuzii/ *Ciel et Espace.*

Kaluga, and Robert Goddard, the engineer, a professor at an equally obscure American college in Massachusetts. They dreamt of using rockets to journey to the planets and the stars. Step by step, they worked out the fundamental physics and many of the details. Gradually, their machines took shape. Ultimately, their dream proved infectious.

In their time, the very idea was considered disreputable, or even a symptom of some obscure derangement. Goddard found that merely mentioning a voyage to other worlds subjected him to ridicule, and he dared not publish or even discuss in public his long-term vision of flights to the stars. As teenagers, both had epiphanal visions of spaceflight that never left them. "I still have dreams in which I fly up to the stars in my machine," Tsiolkovsky wrote in middle age. "It is difficult to work all on your own for many years, in adverse conditions without a gleam of hope, without any help." Many of his contemporaries thought he was truly mad. Those who knew physics better than Tsiolkovsky and Goddard—including *The New York Times* in a dismissive editorial not retracted until the eve of *Apollo 11*—insisted that rockets could not work in a vacuum, that the Moon and the planets were forever beyond human reach.

A generation later, inspired by Tsiolkovsky and Goddard, Wernher von Braun was constructing the first rocket capable of reaching the edge of space, the V-2. But in one of those ironies with which the twentieth century is replete, von Braun was building it for the Nazis—as an instrument of indiscriminate slaughter of civilians, as a "vengeance weapon" for Hitler, the rocket factories staffed with slave labor, untold human suffering exacted in the construction of every booster, and von Braun himself made an officer in the SS. He was aiming at the Moon, he joked unselfconsciously, but hit London instead.

Another generation later, building on the work of Tsiolkovsky and Goddard, extending von Braun's technological genius, we were up there in space, silently circumnavigating the Earth, treading the ancient and desolate lunar surface. Our machines—increasingly competent and autonomous—were spreading through the Solar System, discovering new worlds, ex-

A classic painting by
Chesley Bonestell/
Space Art International of
a V-2 class rocket near its
gantry being prepared
for launch. Courtesy
Frederick C. Durant III.

amining them closely, searching for life, comparing them with Earth.

This is one reason that in the long astronomical perspective there is something truly epochal about "now"—which we can define as the few centuries centered on the year you're reading this book. And there's a second reason: This is the first moment in the history of our planet when any species, by its own voluntary actions, has become a danger to itself—as well as to vast numbers of others. Let me recount the ways:

- We've been burning fossil fuels for hundreds of thousands of years. By the 1960s, there were so many of us burning wood, coal, oil, and natural gas on so large a scale, that scientists began to worry about the increasing greenhouse effect; the dangers of global warming began slowly slipping into public consciousness.
- CFCs were invented in the 1920s and 1930s; in 1974 they were discovered to attack the protective ozone layer. Fifteen years later a worldwide ban on their production was going into effect.
- Nuclear weapons were invented in 1945. It took until 1983 before the global consequences of thermonuclear war were understood. By 1992, large numbers of warheads were being dismantled.
- The first asteroid was discovered in 1801. More or less serious proposals to move them around were floated beginning in the 1980s. Recognition of the potential dangers of asteroid deflection technology followed shortly after.
- Biological warfare has been with us for centuries, but its deadly mating with molecular biology has occurred only lately.
- We humans have already precipitated extinctions of species on a scale unprecedented since the end of the Cretaceous Period. But only in the last decade has the magnitude of these extinctions become clear, and the possibility raised that in our ignorance of the interrelations of life on Earth we may be endangering our own future.

Look at the dates on this list and consider the range of new technologies currently under development. Is it not likely that other dangers of our own making are yet to be discovered, some perhaps even more serious?

In the littered field of discredited self-congratulatory chauvinisms, there is only one that seems to hold up, one sense in which we *are* special: Due to our own actions or inactions, and the misuse of our technology, we live at an extraordinary moment for the Earth at least—the first time that a species has become able to wipe itself out. But this is also, we may note, the first time that a species has become able to journey to the planets and the stars. The two times, brought about by the same technology, coincide— a few centuries in the history of a 4.5-billion-year-old planet. If you were somehow dropped down on the Earth randomly at any moment in the past (or future), the chance of arriving at this critical moment would be less than 1 in 10 million. Our leverage on the future is high just now.

It might be a familiar progression, transpiring on many worlds—a planet, newly formed, placidly revolves around its star; life slowly forms; a kaleidoscopic procession of creatures evolves; intelligence emerges which, at least up to a point, confers enormous survival value; and then technology is invented. It dawns on them that there are such things as laws of Nature, that these laws can be revealed by experiment, and that knowledge of these laws can be made both to save and to take lives, both on unprecedented scales. Science, they recognize, grants immense powers. In a flash, they create world-altering contrivances. Some planetary civilizations see their way through, place limits on what may and what must not be done, and safely pass through the time of perils. Others, not so lucky or so prudent, perish.

Since, in the long run, every planetary society will be endangered by impacts from space, every surviving civilization is obliged to become spacefaring—not because of exploratory or romantic zeal, but for the most practical reason imaginable: staying alive. And once you're out there in space for centuries and millennia, moving little worlds around and engineering planets, your species has been pried loose from its cradle. If they exist, many other civilizations will eventually venture far from home.[*]

A MEANS HAS BEEN OFFERED of estimating how precarious our circumstances are—remarkably, without in any way addressing the nature of the hazards. J. Richard Gott III is an astrophysicist at Princeton University. He asks us to adopt a generalized Copernican principle, something I've described elsewhere as the Principle of Mediocrity. Chances are that we do not live in a truly extraordinary time. Hardly anyone ever did. The probability is high that we're born, live out our days, and die somewhere in the broad middle range of the lifetime of our species (or civilization, or nation). Almost certainly, Gott says, we do not live in first or last times. So if your species is very young, it follows that it's unlikely to last long—because if it *were* to last long, you (and the rest of us alive today) *would* be extraordinary in living, proportionally speaking, so near the beginning.

What then is the projected longevity of our species? Gott concludes, at the 97.5 percent confidence level, that there will be humans for no more than 8 million years. That's his upper limit, about the same as the average lifetime of many mammalian species. In that case, our technology neither harms nor helps. But Gott's lower limit, with the same claimed reliability, is only 12 years. He will not give you 40-to-1 odds that humans will still be

---

[*] Might a planetary civilization that has survived its adolescence wish to encourage others struggling with *their* emerging technologies? Perhaps they would make special efforts to broadcast news of their existence, the triumphant announcement that it's possible to avoid self-annihilation. Or would they at first be very cautious? Having avoided catastrophes of their own making, perhaps they would fear giving away knowledge of their existence, lest some other, unknown, aggrandizing civilization out there in the dark is looking for *Lebensraum* or slavering to put down the potential competition. That might be a reason for us to explore neighboring star systems, but discreetly.

Maybe they would be silent for another reason: because broadcasting the existence of an advanced civilization might encourage emerging civilizations to do less than their best efforts to safeguard their future—hoping instead that someone will come out of the dark and save them from themselves.

around by the time babies now alive become teenagers. In everyday life we try very hard not to take risks so large, not to board airplanes, say, with 1 chance in 40 of crashing. We will agree to surgery in which 95 percent of patients survive only if our disease has a greater than 5 percent chance of killing us. Mere 40-to-1 odds on our species surviving another 12 years would be, if valid, a cause for supreme concern. If Gott is right, not only may we never be out among the stars; there's a fair chance we may not be around long enough even to make the first footfall on another planet.

To me, this argument has a strange, vaporish quality. Knowing nothing about our species except how old it is, we make numerical estimates, claimed to be highly reliable, about its future prospects. How? We go with the winners. Those who have been around are likely to stay around. Newcomers tend to disappear. The only assumption is the quite plausible one that there is nothing special about the moment at which we inquire into the matter. So why is the argument unsatisfying? Is it just that we are appalled by its implications?

Something like the Principle of Mediocrity must have very broad applicability. But we are not so ignorant as to imagine that everything is mediocre. There *is* something special about our time—not just the temporal chauvinism that those who reside in any epoch doubtless feel, but something, as outlined above, clearly unique and strictly relevant to our species' future chances: This is the first time that (a) our exponentiating technology has reached the precipice of self-destruction, but also the first time that (b) we can postpone or avoid destruction by going somewhere else, somewhere off the Earth.

These two clusters of capabilities, (a) and (b), make our time extraordinary in directly contradictory ways—which both (a) strengthen and (b) weaken Gott's argument. I don't know how to predict whether the new destructive technologies will hasten, more than the new spaceflight technologies will delay, human extinction. But since never before have we contrived the means of annihilating ourselves, and never before have we developed the technology for settling other worlds, I think a compelling case can be made that our time is extraordinary precisely in the context of

Gott's argument. If this is true, it significantly increases the margin of error in such estimates of future longevity. The worst is worse, and the best better: Our short-term prospects are even bleaker and—if we can survive the short term—our long-term chances even brighter than Gott calculates.

But the former is no more cause for despair than the latter is for complacency. Nothing forces us to be passive observers, clucking in dismay as our destiny inexorably works itself out. If we cannot quite seize fate by the neck, perhaps we can misdirect it, or mollify it, or escape it.

Of course we must keep our planet habitable—not on a leisurely timescale of centuries or millennia, but urgently, on a timescale of decades or even years. This will involve changes in government, in industry, in ethics, in economics, and in religion. We've never done such a thing before, certainly not on a global scale. It may be too difficult for us. Dangerous technologies may be too widespread. Corruption may be too pervasive. Too many leaders may be focused on the short term rather than the long. There may be too many quarreling ethnic groups, nation-states, and ideologies for the right kind of global change to be instituted. We may be too foolish to perceive even what the real dangers are, or that much of what we hear about them is determined by those with a vested interest in minimizing fundamental change.

However, we humans also have a history of making long-lasting social change that nearly everyone thought impossible. Since our earliest days, we've worked not just for our own advantage but for our children and our grandchildren. My grandparents and parents did so for me. We have often, despite our diversity, despite endemic hatreds, pulled together to face a common enemy. We seem, these days, much more willing to recognize the perils before us than we were even a decade ago. The newly recognized dangers threaten all of us equally. No one can say how it will turn out down here.

THE MOON WAS WHERE the tree of immortality grew in ancient Chinese myth. The tree of longevity if not of immortality, it seems, indeed grows on other worlds. If we were up there among the planets, if there were self-sufficient human communities on many worlds, our species would be insulated from catastrophe. The depletion of the ultraviolet-absorbing shield on one world would, if anything, be a warning to take special care of the shield on another. A cataclysmic impact on one world would likely leave all the others untouched. The more of us beyond the Earth, the greater the diversity of worlds we inhabit, the more varied the planetary engineering, the greater the range of societal standards and values—then the safer the human species will be.

If you grow up living underground in a world with a hundredth of an Earth gravity and black skies through the portals, you have a very different set of perceptions, interests, prejudices, and predispositions than someone who lives on the surface of the home planet. Likewise if you live on the surface of Mars in the throes of terraforming, or Venus, or Titan. This strategy—breaking up into many smaller self-propagating groups, each with somewhat different strengths and concerns, but all marked by local pride—has been widely employed in the evolution of life on Earth, and by our own ancestors in particular. It may, in fact, be key to understanding why we humans are the way we are.[*] This is the second of the missing justifications for a permanent human presence in space: to improve our chances of surviving, not just the catastrophes we can foresee, but also the ones we cannot. Gott also argues that establishing human communities on other worlds may offer us our best chance of beating the odds.

To take out this insurance policy is not very expensive, not on the scale on which we do things on Earth. It would not even require doubling the space budgets of the present spacefaring nations (which, in all cases, are only a small fraction of the military budgets and many voluntary expenditures that might be considered marginal or even frivolous). We could soon be setting humans down on near-Earth asteroids and establishing bases on Mars. We know how to do it, even with present technology, in less than a

---

[*] Cf. *Shadows of Forgotten Ancestors: A Search for Who We Are,* by Carl Sagan and Ann Druyan (New York: Random House, 1992).

human lifetime. And the technologies will quickly improve. We will get better at going into space.

A serious effort to send humans to other worlds is relatively so inexpensive on a *per annum* basis that it cannot seriously compete with urgent social agendas on Earth. If we take this path, streams of images from other worlds will be pouring down on Earth at the speed of light. Virtual reality will make the adventure accessible to millions of stay-on-Earths. Vicarious participation will be much more real than at any earlier age of exploration and discovery. And the more cultures and people it inspires and excites, the more likely it will happen.

But by what right, we might ask ourselves, do we inhabit, alter, and conquer other worlds? If anyone else were living in the Solar System, this would be an important question. If, though, there's no one else in this system but us, don't we have a right to settle it?

Of course, our exploration and homesteading should be enlightened by a respect for planetary environments and the scientific knowledge they hold. This is simple prudence. Of course, exploration and settlement ought to be done equitably and transnationally, by representatives of the entire human species. Our past colonial history is not encouraging in these regards; but this time we are not motivated by gold or spices or slaves or a zeal to convert the heathen to the One True Faith, as were the European explorers of the fifteenth and sixteenth centuries. Indeed, this is one of the chief reasons we're experiencing such intermittent progress, so many fits and starts in the manned space programs of all nations.

Despite all the provincialisms I complained about early in this book, here I find myself an unapologetic human chauvinist. If there were other life in this solar system, it would be in imminent danger because the humans are coming. In such a case, I might even be persuaded that safeguarding our species by settling certain other worlds is offset, in part at least, by the danger we would pose to everybody else. But as nearly as we can tell, so far at least, there is no other life in this system, not one microbe. There's only Earth-life.

In that case, on behalf of Earthlife, I urge that, with full knowledge of our limitations, we vastly increase our knowledge of the Solar System and then begin to settle other worlds.

These are the missing practical arguments: safeguarding the Earth from otherwise inevitable catastrophic impacts and hedging our bets on the many other threats, known and unknown, to the environment that sustains us. Without these arguments, a compelling case for sending humans to Mars and elsewhere might be lacking. But with them—and the buttressing arguments involving science, education, perspective, and hope—I think a strong case can be made. If our long-term survival is at stake, we have a basic responsibility to our species to venture to other worlds.

Sailors on a becalmed sea, we sense the stirring of a breeze.

TIPTOEING THROUGH THE MILKY WAY

I swear by the shelters of the stars (a mighty oath, if you but knew it) . . .
—THE QUR'AN, SURA 56 (7TH CENTURY)

Of course, it is strange to inhabit the earth no longer,
To give up customs one barely had time to learn . . .
—RAINER MARIA RILKE, "THE FIRST ELEGY" (1923)

The prospect of scaling heaven, of ascending to the sky, of altering other worlds to suit our purposes—no matter how well intentioned we may be—sets the warning flags flying: We remember the human inclination toward overweening pride; we recall our fallibility and misjudgments when presented with powerful new technologies. We recollect the story of the Tower of Babel, a building "whose top may reach unto heaven," and God's fear about our species, that now "nothing will be restrained from them which they have imagined to do."

OPPOSITE: Having crossed the light years over many generations, an asteroidal habitat arrives at the Earthlike planet of another star. Painting © David A. Hardy.

We come upon Psalm 15, which stakes a divine claim to other worlds: "[T]he heavens are the Lord's, but the Earth hath he given to the children of men." Or Plato's retelling of the Greek analogue of Babel—the tale of Otys and Ephialtes. They were mortals who "dared to scale heaven." The gods were faced with a choice. Should they kill the upstart humans "and annihilate [their]

The best portrait of the Milky Way Galaxy according to present knowledge. This view is from a point almost 60,000 light-years from the galactic center, and about 10,000 light-years above the plane of the Galaxy. We are so far away that only the brightest stars and nebulae can be seen. The Sun is on the outskirts of the Sagittarius spiral arm—at the center of the picture and halfway down from the galactic center. Painting by Jon Lomberg. © 1992 Jon Lomberg and National Air and Space Museum. A 40" X 28" poster of this painting can be obtained from The Planetary Society at the address listed in the Acknowledgments.

race with thunderbolts"? On the one hand, "this would be the end of the sacrifices and worship which men offered" the gods and which gods craved. "But, on the other hand, the gods could not suffer [such] insolence to be unrestrained."

If, in the long term, though, we have no alternative, if our choice really is many worlds or none, we are in need of other sorts of myth, myths of encouragement. They exist. Many religions, from Hinduism to Gnostic Christianity to Mormon doctrine, teach that—as impious as it may sound—it is the goal of humans to *become* gods. Or consider a story in the Jewish Talmud left out of the Book of Genesis. (It is in doubtful accord with the account of the apple, the Tree of Knowledge, the Fall, and the expulsion from Eden.) In the Garden, God tells Eve and Adam that He has intentionally left the Universe unfinished. It is the responsibility of humans, over countless generations, to participate with God in a "glorious" experiment—"completing the Creation."

The burden of such a responsibility is heavy, especially on so weak and imperfect a species as ours, one with so unhappy a history. Nothing remotely like "completion" can be attempted without vastly more knowledge than we have today. But perhaps, if our very existence is at stake, we will find ourselves able to rise to this supreme challenge.

ALTHOUGH HE DID NOT quite use any of the arguments of the preceding chapter, it was Robert Goddard's intuition that "the navigation of interplanetary space must be effected to ensure the continuance of the race." Konstantin Tsiolkovsky made a similar judgment:

> There are countless planets, like many island Earths . . . Man occupies one of them. But why could he not avail himself of others, and of the might of numberless suns? . . . When the Sun has exhausted its energy, it would be logical to leave it and look for another, newly kindled, star still in its prime.

This might be done earlier, he suggested, long before the Sun dies, "by adventurous souls seeking fresh worlds to conquer."

But as I rethink this whole argument, I'm troubled. Is it too much Buck Rogers? Does it demand an absurd confidence in future technology? Does it ignore my own admonitions about human fallibility? Surely in the short term it's biased against technologically less-developed nations. Are there no practical alternatives that avoid these pitfalls?

All our self-inflicted environmental problems, all our weapons of mass destruction are products of science and technology. So, you might say, let's just back off from science and technology. Let's admit that these tools are simply too hot to handle. Let's create a simpler society, in which no matter how careless or short-sighted we are, we're incapable of altering the environment on a global or even on a regional scale. Let's throttle back to a minimal, agriculturally intensive technology, with stringent controls on new knowledge. An authoritarian theocracy is a tried-and-true way to enforce the controls.

Such a world culture is unstable, though, in the long run if not the short—because of the speed of technological advance. Human propensities for self-betterment, envy, and competition will always be throbbing subsurface; opportunities for short-term, local advantage will sooner or later be seized. Unless there are severe constraints on thought and action, in a flash we'll be back to where we are today. So controlled a society must grant great powers to the elite that does the controlling, inviting flagrant abuse and eventual rebellion. It's very hard—once we've seen the riches, conveniences, and lifesaving medicines that technology offers—to squelch human inventiveness and acquisitiveness. And while such a devolution of the global civilization, were it possible, might conceivably address the problem of self-inflicted technological catastrophe, it would also leave us defenseless against eventual asteroidal and cometary impacts.

Or you might imagine throttling back much further, back to hunter-gatherer society, where we live off the natural products of the land and abandon even agriculture. Javelin, digging stick, bow, arrow, and fire would then be technology enough. But the Earth could support at the very most a few tens of millions of hunter-gatherers. How could we get down to such low population levels without instigating the very catastrophes we are trying to avoid? Besides, we hardly know how to live the hunter-gatherer life anymore: We've forgotten their cultures, their skills, their tool-kits. We've killed off almost all of them, and we've destroyed much of the environment that sustained them. Except for a tiny remnant of us, we might not be able, even if we gave it high priority, to go back. And again, even if we could return, we would be helpless before the impact catastrophe that inexorably will come.

The alternatives seem worse than cruel: They are ineffective.

Many of the dangers we face indeed arise from science and technology—but, more fundamentally, because we have become powerful without becoming commensurately wise. The world-altering powers that technology has delivered into our hands now require a degree of consideration and foresight that has never before been asked of us.

Science cuts two ways, of course; its products can be used for both good and evil. But there's no turning back from science. The early warnings about technological dangers also come from science. The solutions may well require more of us than just a technological fix. Many will have to become scientifically literate. We may have to change institutions and behavior. But our problems, whatever their origin, cannot be solved apart from science. The technologies that threaten us and the circumvention of those threats both issue from the same font. They are racing neck and neck.

In contrast, with human societies on several worlds, our prospects would be far more favorable. Our portfolio would be diversified. Our eggs would be, almost literally, in many baskets. Each society would tend to be proud of the virtues of its world, its planetary engineering, its social conventions, its hereditary predispositions. Necessarily, cultural differences would be cherished and exaggerated. This diversity would serve as a tool of survival.

When the off-Earth settlements are better able to fend for themselves, they will have every reason to encourage technological advance, openness of spirit, and adventure—even if those left on Earth are obliged to prize caution, fear new knowledge, and institute Draconian social controls. After the first few self-sustaining communities are established on other worlds, the Earthlings might also be able to relax their strictures and lighten up. The humans in space would provide those on Earth with real protection against rare but catastrophic collisions by asteroids or comets on rogue trajectories. Of course, for this very reason, humans in space would hold the upper hand in any serious dispute with those on Earth.

The prospects of such a time contrast provocatively with forecasts that the progress of science and technology is now near some asymptotic limit; that art, literature, and music are never to approach, much less exceed, the heights our species has, on occasion, already touched; and that political life on Earth is about to settle into some rock-stable liberal democratic world government, identified, after Hegel, as "the end of history." Such an expansion into space also contrasts with a different but likewise discernible trend in recent times—toward authoritarianism, censorship, ethnic hatred, and a deep suspicion of curiosity and learning. Instead, I think that, after some debugging, the settlement of the Solar System presages an open-ended era of dazzling advances in science and technology; cultural flowering; and wide-ranging experiments, up there in the sky, in government and social organization. In more than one respect, exploring the Solar System and homesteading other worlds constitutes the beginning, much more than the end, of history.

IT'S IMPOSSIBLE, for us humans at least, to look into our future, certainly not centuries ahead. No one has ever done so with any consistency and detail. I certainly do not imagine that I can. I have, with some trepidation, gone as far as I have to this point in the book, because we are just recognizing the truly unprecedented challenges brought on by our technology. These challenges have, I think, occasional straightforward implications, some of which I've tried briefly to lay out. There are also less straightforward, much longer-term implications about which I'm even less confident. Nevertheless, I'd like to present them too for your consideration:

Even when our descendants are established on near-Earth asteroids and Mars and the moons of the outer Solar System and the Kuiper Comet Belt, it still won't be entirely safe. In the long run, the Sun may generate stupendous X-ray and ultraviolet outbursts; the Solar System will enter one of the vast interstellar clouds lurking nearby and the planets will darken and cool; a shower of deadly comets will come roaring out of the Oort Cloud threatening civilizations on many adjacent worlds; we will recognize that a nearby star is about to become a supernova. In the *really* long run, the Sun—on its way to becoming a red giant star—will get bigger and brighter, the Earth will begin to lose its air and water to space, the soil will char, the oceans will evaporate and boil, the rocks will

The finest ground-based
telescopic image so far of
the Great Nebula in Orion—
a spawning ground for stars
1500 light-years away.
Courtesy ROE/Anglo-Australian
Observatory. Photograph by
David Malin.

vaporize, and our planet may even be swallowed up into the interior of the Sun.

Far from being made for us, eventually the Solar System will become too dangerous for us. In the long run, putting all our eggs in a single stellar basket, no matter how reliable the Solar System has been lately, may be too risky. In the long run, as Tsiolkovsky and Goddard long ago recognized, we need to leave the Solar System.

If that's true for us, you might very well ask, why isn't it true for others? And if it *is* true for others, why aren't they here? There are many possible answers, including the contention that they *have* come here—although the evidence for that is pitifully slim. Or there may be no one else out there, because they destroy themselves, with almost no exceptions, before they achieve interstellar flight; or because in a galaxy of 400 billion suns ours is the first technical civilization.

A more likely explanation, I think, issues from the simple fact that space is vast and the stars are far apart. Even if there were civilizations much older and more advanced than we—expanding out from their home worlds, reworking new worlds, and then continuing onward to other stars—they would be unlikely, according to calculations performed by William I. Newman of UCLA and me, to be here. Yet. And because the speed of light is

Close-up of the Orion Nebula from the Hubble Space Telescope. This immense cloud of gas is illuminated by bright, hot, and very young stars near the bottom of the picture. Strewn across center are a number of cocoon-shaped objects. These are young stars, only a few hundred thousand years old, surrounded by disks of dust and gas roughly the size of our solar system. Courtesy C. R. O'Dell/Rice University/NASA.

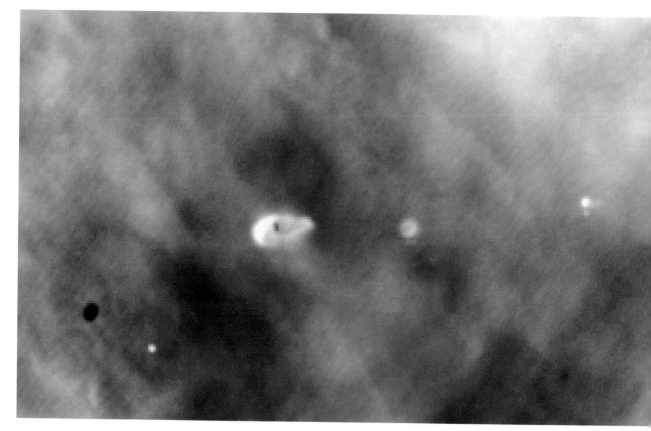

Close-up of a few of the pre-planetary clouds of gas and dust in the Orion Nebula. Of the 110 young stars examined here, disks have been found around 56 of them. It is much easier to detect the star than the disk, so it is possible that *all* the young stars have pre-planetary disks around them. The implication is clear: Many, perhaps most, and maybe even all mature stars may have planetary systems. This is an encouragement to aspiring interstellar space travelers. Courtesy C. R. O'Dell/Rice University/NASA.

finite, the TV and radar news that a technical civilization has arisen on some planet of the Sun has not reached them. Yet.

Should optimistic estimates prevail and one in every million stars shelters a nearby technological civilization, and if as well they're randomly strewn through the Milky Way—were these provisos to hold—then the nearest one, we recall, would be a few hundred light-years distant: at the closest, maybe 100 light-years, more likely a thousand light-years—and, of course, perhaps nowhere, no matter how far. Suppose the nearest civilization on a planet of another star is, say, 200 light-years away. Then, some 150 years from now they'll begin to receive our feeble post–World War II television and radar emission. What will they make of it? With each passing year the signal will get louder, more interesting, perhaps more alarming. Eventually, they may respond: by returning a radio message, or by visiting. In either case, the response will likely be limited by the finite value of the speed of light. With these wildly uncertain numbers, the answer to our unintentional mid-century call into the depths of space will not arrive until around the year 2350. If they're farther away, of course, it will take longer; and if much farther away, much longer. The interesting possibility arises that our first receipt of a message from an alien civilization, a message intended for us (not just an all-points bulletin), will occur in a time when we are well situated on many worlds in our solar system and preparing to move on.

With or without such a message, though, we will have reason to continue outward, seeking other solar systems. Or—even safer in this unpredictable and violent sector of the Galaxy—to sequester some of us in self-sufficient habitations in interstellar space, far from the dangers constituted by the stars. Such a future would, I think, naturally evolve, by slow increments, even without any grand goal of interstellar travel:

For safety, some communities may wish to sever their ties with the rest of humanity—uninfluenced by other societies, other ethical codes, other technological imperatives. In a time when comets and asteroids are being routinely repositioned, we will be able to populate a small world and then cut it loose. In successive generations, as this world sped outward, the Earth would fade from bright star to pale dot to invisibility; the Sun would appear dimmer, until it was no more than a vaguely yellow point of light, lost among thousands of others. The travelers would approach interstellar night. Some such communities may be content with occasional radio and laser traffic with the old home worlds. Others, confident of the superiority of their own survival chances and wary of contamination, may try to disappear. Perhaps all contact with them will ultimately be lost, their very existence forgotten.

Even the resources of a sizable asteroid or comet are finite, though, and eventually more resources must be sought else-where—especially water, needed for drink, for a breathable oxygen atmosphere, and for hydrogen to power fusion reactors. So in the long run these communities must migrate from world to world, with no lasting loyalty to any. We might call it "pioneering," or "homesteading." A less sympathetic observer might describe it as sucking dry the resources of little world after little world. But there are a trillion little worlds in the Oort Comet Cloud.

Living in small numbers on a modest stepmother world far from the Sun, we will know that every scrap of food and every drop of water is dependent on the smooth operation of a far-sighted technology; but these conditions are not radically unlike those to which we are already accustomed. Digging resources out of the ground and stalking passing resources seem oddly familiar, like a forgotten memory of childhood: It is, with a few significant changes, the strategy of our hunter-gatherer ancestors. For 99.9 percent of the tenure of humans on Earth, we lived such a life. Judging from some of the last surviving hunter-gatherers just before they were engulfed by the present global civilization, we may have been relatively happy. It's the kind of life that forged us. So after a brief, only partially successful sedentary experiment, we may become wanderers again—more technological than last time, but even then our technology, stone tools and fire, was our only hedge against extinction.

If safety lies in isolation and remoteness, then some of our descendants will eventually emigrate to the outer comets of the Oort Cloud. With a trillion cometary nuclei, each separated from the

A new Earth circles a Sun-like star in a widely separated binary star system. Painting by Don Davis.

next by about as much as Mars is from Earth, there will be a great deal to do out there.*

The outer edge of the Sun's Oort Cloud is perhaps halfway to the nearest star. Not every other star has an Oort Cloud, but many probably do. As the Sun passes nearby stars, our Oort Cloud will encounter, and partially pass through, other comet clouds, like two swarms of gnats interpenetrating but not colliding. To occupy a comet of another star will then be not much more difficult than to occupy one of our own. From the frontiers of some other solar system the children of the blue dot may peer longingly at the moving points of light denoting substantial (and well-lit) planets. Some communities—feeling the ancient human love for oceans and sunlight stirring within them—may begin the long journey down to the bright, warm, and clement planets of a new sun.

Other communities may consider this last strategy a weakness. Planets are associated with natural catastrophes. Planets may have pre-existing life and intelligence. Planets are easy for other beings to find. Better to remain in the darkness. Better to spread ourselves among many small and obscure worlds. Better to stay hidden.

ONCE WE CAN SEND our machines and ourselves far from home, far from the planets—once we really enter the theater of the Universe—we are bound to come upon phenomena unlike anything we've ever encountered. Here are three possible examples:

First: Starting some 550 astronomical units (AU) out—about ten times farther from the Sun than Jupiter, and therefore much more accessible than the Oort Cloud—there's something extraordinary. Just as an ordinary lens focuses far-off images, so does gravity. (Gravitational lensing by distant stars and galaxies is now being detected.) Five hundred fifty AU from the Sun—only a year away

---

* Even if we are not in any particular hurry, we may be able by then to make small worlds move faster than we can make spacecraft move today. If so, our descendants will eventually overtake the two *Voyager* spacecraft—launched in the remote twentieth century—before they leave the Oort Cloud, before they make for interstellar space. Perhaps they will retrieve these derelict ships of long ago. Or perhaps they will permit them to sail on.

if we could travel at 1 percent the speed of light—is where the focus begins (although when effects of the solar corona, the halo of ionized gas surrounding the Sun, are taken into account, the focus may be considerably farther out). There, distant radio signals are enormously enhanced, amplifying whispers. The magnification of distant images would allow us (with a modest radio telescope) to resolve a continent at the distance of the nearest star and the inner Solar System at the distance of the nearest spiral galaxy. If you are free to roam an imaginary spherical shell at the appropriate focal distance and centered on the Sun, you are free to explore the Universe in stupendous magnification, to peer at it with unprecedented clarity, to eavesdrop on the radio signals of distant civilizations, if any, and to glimpse the earliest events in the history of the Universe. Alternatively, the lens could be used the other way, to amplify a very modest signal of ours so it could be heard over immense distances. There are reasons that draw us to hundreds and thousands of AU. Other civilizations will have their own regions of gravitational focusing, depending on the mass and radius of their star, some a little closer, some a little farther away than ours. Gravitational lensing may serve as a common inducement for civilizations to explore the regions just beyond the planetary parts of their solar systems.

Second: Spend a moment thinking about brown dwarfs, hypothetical very low temperature stars, considerably more massive than Jupiter, but considerably less massive than the Sun. Nobody knows if brown dwarfs exist. Some experts, using nearer stars as gravitational lenses to detect the presence of more distant ones, claim to have found evidence of brown dwarfs. From the tiny fraction of the whole sky that has so far been observed by this technique, an enormous number of brown dwarfs is inferred. Others disagree. In the 1950s, it was suggested by the astronomer Harlow Shapley of Harvard that brown dwarfs—he called them "Lilliputian stars"—were inhabited. He pictured their surfaces as warm as a June day in Cambridge, with lots of area. They would be stars that humans could survive on and explore.

Third: The physicists B. J. Carr and Stephen Hawking of Cambridge University have shown that fluctuations in the density

A brown dwarf, a hypothetical very cool star that some astronomers think may be abundant in interstellar space. Earthlike temperatures would prevail on the surfaces of some of them. Painting by Michael Carroll.

of matter in the earliest stages of the Universe could have generated a wide variety of small black holes. Primordial black holes—if they exist—must decay by emitting radiation to space, a consequence of the laws of quantum mechanics. The less massive the black hole, the faster it dissipates. Any primordial black hole in the final stages of decay today would have to weigh about as much as a mountain. All the smaller ones are gone. Since the abundance—to say nothing of the existence—of primordial black holes depends on what happened in the earliest moments after the Big Bang, no one can be sure that there are any to be found; we certainly can't be sure that any lie nearby. Not very restrictive upper limits on their abundance have been set by the failure so far to find short gamma ray pulses, a component of the Hawking radiation.

In a separate study, G.E. Brown of Caltech and the pioneering nuclear physicist Hans Bethe of Cornell suggest that about a billion *non*primordial black holes are strewn through the Galaxy, generated in the evolution of stars. If so, the nearest may be only 10 or 20 light-years away.

If there are black holes within reach—whether they're as massive as mountains or as stars—we will have amazing physics to study firsthand, as well as a formidable new source of energy. By no means do I claim that brown dwarfs or primordial black holes are likely within a few light-years, or anywhere. But as we enter interstellar space, it is inevitable that we will stumble upon whole new categories of wonders and delights, some with transforming practical applications.

I do not know where my train of argument ends. As more time passes, attractive new denizens of the cosmic zoo will draw us farther outward, and increasingly improbable and deadly catastrophes must come to pass. The probabilities are cumulative. But, as time goes on, technological species will also accrue greater and greater powers, far surpassing any we can imagine today. Perhaps, if we are very skillful (lucky, I think, won't be enough), we will ultimately spread far from home, sailing through the starry archipelagos of the vast Milky Way Galaxy. If we come upon anyone else—or, more likely, if they come upon us—we will harmoniously interact. Since other spacefaring civilizations are likely to be much more advanced than we, quarrelsome humans in interstellar space are unlikely to last long.

Eventually, our future may be as Voltaire, of all people, imagined:

> Sometimes by the help of a sunbeam, and sometimes by the convenience of a comet, [they] glided from sphere to sphere, as a bird hops from bough to bough. In a very little time [they] posted through the Milky Way . . .

We are, even now, discovering vast numbers of gas and dust disks around young stars—the very structures out of which, in our solar system four and a half billion years ago, the Earth and the other planets formed. We're beginning to understand how fine dust grains slowly grow into worlds; how big Earthlike planets accrete and then quickly capture hydrogen and helium to become the hidden cores of gas giants; and how small terrestrial planets remain comparatively bare of atmosphere. We are reconstructing the histories of worlds—how mainly ices and organics collected together in the chilly outskirts of the early Solar System, and mainly rock and metal in the inner regions warmed by the young Sun. We have begun to recognize the dominant role of early collisions in knocking worlds over, gouging huge craters and basins in their surfaces and interiors, spinning them up, making and obliterating moons, creating rings, carrying, it may be, whole oceans down from the skies, and then depositing a veneer of organic matter as the neat finishing touch in the creation of worlds. We are beginning to apply this knowledge to other systems.

In the next few decades we have a real chance of examining the layout and something of the composition of many other mature planetary systems around nearby stars. We will begin to know which aspects of our system are the rule and which the exception. What is more common—planets like Jupiter, planets like Neptune, or planets like Earth? Or do all other systems have Jupiters and Neptunes and Earths? What other categories of worlds are there, currently unknown to us? Are all solar systems embedded in a vast spherical cloud of comets? Most stars in the sky are not solitary suns like our own, but double or multiple systems in which the stars are in mutual orbit. Are there planets in such systems? If so, what are they like? If, as we now think, planetary systems are a routine consequence of the origin of suns, have they followed very different evolutionary paths elsewhere? What do elderly planetary systems, billions of years more evolved than ours, look like? In the next few centuries our knowledge of other systems will become increasingly comprehensive. We will begin to know which to visit, which to seed, and which to settle.

Imagine we could accelerate continuously at 1 g—what we're comfortable with on good old *terra firma*—to the midpoint of our voyage, and decelerate continuously at 1 g until we arrive at our destination. It would then take a day to get to Mars, a week and a half to Pluto, a year to the Oort Cloud, and a few years to the nearest stars.

Even a modest extrapolation of our recent advances in transportation suggests that in only a few centuries we will be able to travel close to the speed of light. Perhaps this is hopelessly optimistic. Perhaps it will really take millennia or more. But unless we destroy ourselves first we will be inventing new technologies as strange to us as *Voyager* might be to our hunter-gatherer ancestors. Even today we can think of ways—clumsy, ruinously expensive,

Speed of human transportation over the last few centuries. In the twentieth century we have made an improvement by a factor of 3,000 or so, from a little more than 10 miles per hour for early automobiles to something like thirty thousand miles per hour for the *Voyager* spacecraft. How will this trend continue?

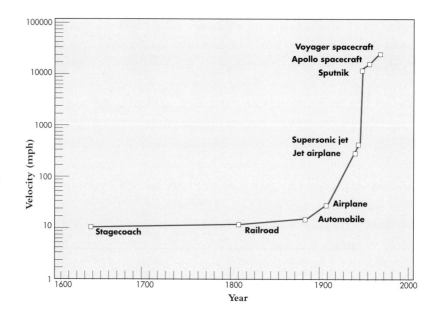

Extrapolation of recent trends in the speed of transportation. The vertical axis goes from one mile per hour to a billion miles per hour, 50 percent more than the speed of light. While taking any particular extrapolation seriously would be foolhardy, the three extrapolations shown here at least suggest that human technology may be approaching the speed of light in the next few centuries, or perhaps even sooner.

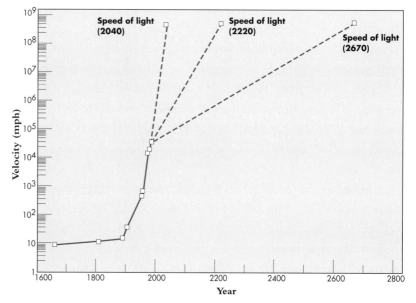

inefficient to be sure—of constructing a starship that approaches light speed. In time, the designs will become more elegant, more affordable, more efficient. The day will come when we overcome the necessity of jumping from comet to comet. We will begin to soar through the light-years and, as St. Augustine said of the gods of the ancient Greeks and Romans, colonize the sky.

Such descendants may be tens or hundreds of generations re-moved from anyone who ever lived on the surface of a planet. Their cultures will be different, their technologies far advanced, their languages changed, their association with machine intelli-gence much more intimate, perhaps their very appearance markedly altered from that of their nearly mythical ancestors who first tentatively set forth in the late twentieth century into the sea of space. But they will be human, at least in large part; they will be practitioners of high technology; they will have historical records. Despite Augustine's judgment on Lot's wife, that "no one who is being saved should long for what he is leaving," they will not wholly forget the Earth.

But we're not nearly ready, you may be thinking. As Voltaire put it in his *Memnon,* "our little terraqueous globe is the mad-house of those hundred thousand millions* of worlds." We, who cannot even put our own planetary home in order, riven with ri-valries and hatreds, despoiling our environment, murdering one another through irritation and inattention as well as on deadly purpose, and moreover a species that until only recently was con-vinced that the Universe was made for its sole benefit—are we to venture out into space, move worlds, re-engineer planets, spread to neighboring star systems?

I do not imagine that it is precisely *we,* with our present cus-toms and social conventions, who will be out there. If we continue to accumulate only power and not wisdom, we will surely destroy ourselves. Our very existence in that distant time requires that we will have changed our institutions and ourselves. How can I dare to guess about humans in the far future? It is, I think, only a matter of natural selection. If we become even slightly more violent, shortsighted, ignorant, and selfish than we are now, almost cer-tainly we will have no future.

If you're young, it's just possible that we will be taking our first steps on near-Earth asteroids and Mars during your lifetime. To spread out to the moons of the Jovian planets and the Kuiper Comet Belt will take many generations more. The Oort Cloud

---

* A value that nicely approximates modern estimates of the number of planets orbiting stars in the Milky Way Galaxy.

will require much longer still. By the time we're ready to settle even the nearest other planetary systems, we will have changed. The simple passage of so many generations will have changed us. The different circumstances we will be living under will have changed us. Prostheses and genetic engineering will have changed us. Necessity will have changed us. We're an adaptable species.

It will not be we who reach Alpha Centauri and the other nearby stars. It will be a species very like us, but with more of our strengths and fewer of our weaknesses, a species returned to circumstances more like those for which it was originally evolved, more confident, farseeing, capable, and prudent—the sorts of beings we would want to represent us in a Universe that, for all we know, is filled with species much older, much more powerful, and very different.

The vast distances that separate the stars are providential. Beings and worlds are quarantined from one another. The quarantine is lifted only for those with sufficient self-knowledge and judgment to have safely traveled from star to star.

A habitable, almost Earthlike world orbits a close-in Jovian planet of a nearby star. Painting by Kasuaki Iwasaki.

An ion starship capable of traveling close to light speed arrives at a habitable planet of a nearby star. Painting © David A. Hardy.

ON IMMENSE TIMESCALES, in hundreds of millions to billions of years, the centers of galaxies explode. We see, scattered across deep space, galaxies with "active nuclei," quasars, galaxies distorted by collisions, their spiral arms disrupted, star systems blasted with radiation or gobbled up by black holes—and we gather that on such timescales even interstellar space, even galaxies may not be safe.

There is a halo of dark matter surrounding the Milky Way, extending perhaps halfway to the distance of the next spiral galaxy (M31 in the constellation Andromeda, which also contains hundreds of billions of stars). We do not know what this dark matter is, or how it is arranged—but some* of it may be in worlds un-

---

* Most of it may be in "nonbaryonic" matter, not made of our familiar protons and neutrons, and not anti-matter either. Over 90 percent of the mass of the Universe seems to be in this dark, quintessential, deeply mysterious stuff wholly unknown on Earth. Perhaps we will one day not only understand it, but also find a use for it.

ABOVE: The Large Magellanic Cloud (LMC), a satellite galaxy of the Milky Way, about 170,000 light-years distant. Courtesy ROE/Anglo-Australian Observatory. Photograph by David Malin. BELOW: In the night sky of a planet in the LMC, the Milky Way rises. Painting by Michael Carroll.

tethered to individual stars. If so, our descendants of the remote future will have an opportunity, over unimaginable intervals of time, to become established in intergalactic space, and to tiptoe to other galaxies.

But on the timescale for populating our galaxy, if not long before, we must ask: How immutable is this longing for safety that drives us outward? Will we one day feel content with the time our species has had and our successes, and willingly exit the cosmic stage? Millions of years from now—probably much sooner—we will have made ourselves into something else. Even if we do nothing intentionally, the natural process of mutation and selection will have worked our extinction or evolved us into some other species on just such a timescale (if we may judge by other mammals). Over the typical lifetime of a mammalian species, even if we were able to travel close to the speed of light and were dedicated to nothing else, we could not, I think, explore even a representative fraction of the Milky Way Galaxy. There's just too much of it. And beyond are a hundred billion galaxies more. Will our present motivations remain unchanged over geological, much less cosmological, timescales—when we ourselves have been transfigured? In

M31, the great galaxy in the constellation Andromeda (with one of its satellite galaxies), seen through the foreground of stars in the Milky Way. M31 is about 2.2 million light-years away. Courtesy Bill and Sally Fletcher.

NGC 3628, a spiral galaxy seen edge-on. Courtesy Anglo-Australian Observatory. Photograph by David Malin.

The peculiar elliptical galaxy Centaurus A, 14 million light-years away. Courtesy Anglo-Australian Observatory. Photograph by David Malin.

such remote epochs, we may discover outlets for our ambitions far grander and more worthy than merely populating an unlimited number of worlds.

Perhaps, some scientists have imagined, we will one day create new forms of life, link minds, colonize stars, reconfigure galaxies, or prevent, in a nearby volume of space, the expansion of the Universe. In a 1993 article in the journal *Nuclear Physics,* the physicist Andrei Linde—conceivably, in a playful mood—suggests that laboratory experiments (it would have to be quite a laboratory) to create separate, closed-off, expanding universes might ultimately be possible. "However," he writes to me, "I myself do not know whether [this suggestion] is simply a joke or something else." In such a list of projects for the far future, we will have no difficulty in recognizing a continuing human ambition to arrogate powers once considered godlike—or, in that other more encouraging metaphor, to complete the Creation.

FOR MANY PAGES NOW, we have left the realm of plausible conjecture for the heady intoxication of nearly unconstrained speculation. It is time to return to our own age.

My grandfather, born before radio waves were even a laboratory curiosity, almost lived to see the first artificial satellite beeping down at us from space. There are people who were born before there was such a thing as an airplane, and who in old age saw four ships launched to the stars. For all our failings, despite our limitations and fallibilities, we humans are capable of greatness. This is true of our science and some areas of our technology, of our art, music, literature, altruism, and compassion, and even, on rare occasion, of our statecraft. What new wonders undreamt of in our time will we have wrought in another generation? And another? How far will our nomadic species have wandered by the end of the next century? And the next millennium?

Two billion years ago our ancestors were microbes; a half-billion years ago, fish; a hundred million years ago, something like mice; ten million years ago, arboreal apes; and a million years ago, proto-humans puzzling out the taming of fire. Our evolutionary lineage is marked by mastery of change. In our time, the pace is quickening.

When we first venture to a near-Earth asteroid, we will have entered a habitat that may engage our species forever. The first voyage of men and women to Mars is the key step in transforming us into a multiplanet species. These events are as momentous as the colonization of the land by our amphibian ancestors and the descent from the trees by our primate ancestors.

Fish with rudimentary lungs and fins slightly adapted for walking must have died in great numbers before establishing a permanent foothold on the land. As the forests slowly receded, our upright apelike forebears often scurried back into the trees, fleeing the predators that stalked the savannahs. The transitions were painful, took millions of years, and were imperceptible to those involved. In our case the transition occupies only a few generations, and with only a handful of lives lost. The pace is so swift that we are still barely able to grasp what is happening.

Once the first children are born off Earth; once we have bases and homesteads on asteroids, comets, moons, and planets; once we're living off the land and bringing up new generations on other worlds, something will have changed forever in human history. But inhabiting other worlds does not imply abandoning this one, any more than the evolution of amphibians meant the end of fish. For a very long time only a small fraction of us will be out there.

"In modern Western society," writes the scholar Charles Lindholm,

> the erosion of tradition and the collapse of accepted religious belief leaves us without a *telos* [an end to which we strive], a sanctified notion of humanity's potential. Bereft of a sacred project, we have only a demystified image of a frail and fallible humanity no longer capable of becoming god-like.

I believe it is healthy—indeed, essential—to keep our frailty and fallibility firmly in mind. I worry about people who aspire to be "god-like." But as for a long-term goal and a sacred project, there is one before us. On it the very survival of our species depends. If we have been locked and bolted into a prison of the self, here is an escape hatch—something worthy, something vastly larger than

FOLLOWING PAGE:
Courtesy ROE/Anglo-Australian Observatory. Photograph by David Malin.

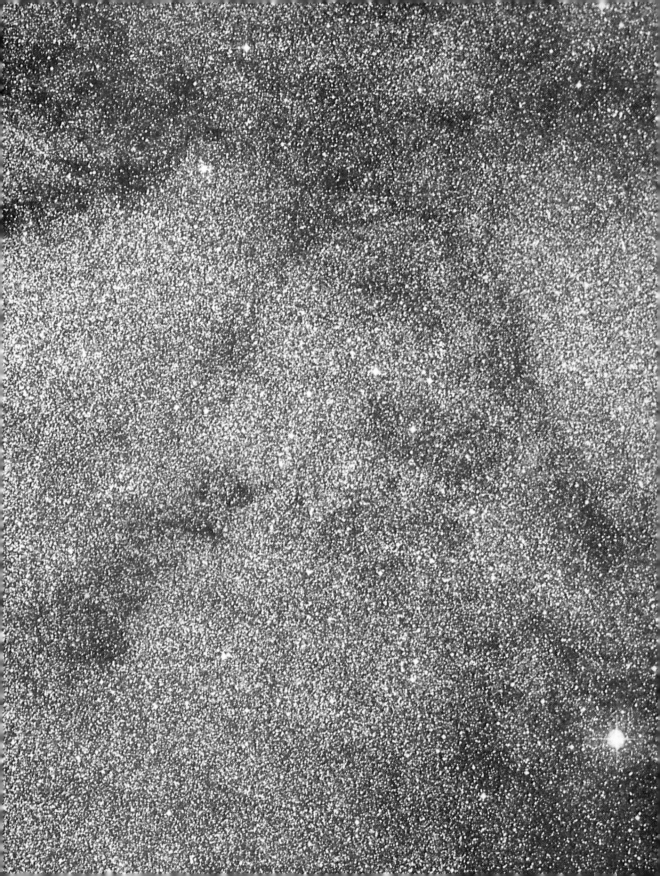

ourselves, a crucial act on behalf of humanity. Peopling other worlds unifies nations and ethnic groups, binds the generations, and requires us to be both smart and wise. It liberates our nature and, in part, returns us to our beginnings. Even now, this new *telos* is within our grasp.

The pioneering psychologist William James called religion a "feeling of being at home in the Universe." Our tendency has been, as I described in the early chapters of this book, to pretend that the Universe is how we wish our home would be, rather than to revise our notion of what's homey so it embraces the Universe. If, in considering James' definition, we mean the *real* Universe, then we have no true religion yet. That is for another time, when the sting of the Great Demotions is well behind us, when we are acclimatized to other worlds and they to us, when we are spreading outward to the stars.

The Cosmos extends, for all practical purposes, forever. After a brief sedentary hiatus, we are resuming our ancient nomadic way of life. Our remote descendants, safely arrayed on many worlds through the Solar System and beyond, will be unified by their common heritage, by their regard for their home planet, and by the knowledge that, whatever other life may be, the only humans in all the Universe come from Earth.

They will gaze up and strain to find the blue dot in their skies. They will love it no less for its obscurity and fragility. They will marvel at how vulnerable the repository of all our potential once was, how perilous our infancy, how humble our beginnings, how many rivers we had to cross before we found our way.

REFERENCES

(a few citations and suggestions for further reading)

PLANETARY EXPLORATION IN GENERAL:

J. Kelly Beatty and Andrew Chaiken, editors, *The New Solar System*, third edition (Cambridge: Cambridge University Press, 1990).

Eric Chaisson and Steve McMillan, *Astronomy Today* (Englewood Cliffs, NJ: Prentice Hall, 1993).

Esther C. Goddard, editor, *The Papers of Robert H. Goddard* (New York: McGraw-Hill, 1970) (three volumes).

Ronald Greeley, *Planetary Landscapes*, second edition (New York: Chapman and Hall, 1994).

William J. Kaufmann III, *Universe,* fourth edition (New York: W. H. Freeman, 1993).

Harry Y. McSween, Jr., *Stardust to Planets* (New York: St. Martin's, 1994).

Ron Miller and William K. Hartmann, *The Grand Tour: A Traveler's Guide to the Solar System,* revised edition (New York: Workman, 1993).

David Morrison, *Exploring Planetary Worlds* (New York: Scientific American Books, 1993).

Bruce C. Murray, *Journey to the Planets* (New York: W. W. Norton, 1989).

Jay M. Pasachoff, *Astronomy: From Earth to the Universe* (New York: Saunders, 1993).

Carl Sagan, *Cosmos* (New York: Random House, 1980).

Konstantin Tsiolkovsky, *The Call of the Cosmos* (Moscow: Foreign Languages Publishing House, 1960) (English translation).

CHAPTER 3, THE GREAT DEMOTIONS

John D. Barrow and Frank J. Tipler, *The Anthropic Cosmological Principle* (New York: Oxford University Press, 1986).

A. Linde, *Particle Physics and Inflationary Cosmology* (Harwood Academy Publishers, 1991).

B. Stewart, "Science or Animism?," *Creation/Evolution,* vol. 12, no. 1 (1992), pp. 18–19.

Steven Weinberg, *Dreams of a Final Theory* (New York: Vintage Books, 1994).

CHAPTER 4, A UNIVERSE NOT MADE FOR US

Bryan Appleyard, *Understanding the Present: Science and the Soul of Modern Man* (London: Picador/Pan Books Ltd. 1992). Passages quoted appear, in order, on the following pages: 232, 27, 32, 19, 19, 27, 9, xiv, 137, 112-113, 206, 10, 239, 8, 8.

J. B. Bury, *History of the Papacy in the 19th Century* (New York: Schocken, 1964). Here, as in many other sources, the 1864 *Syllabus* is transcribed into its "positive" form (e.g., "Divine revelation is perfect") rather than as part of a list of condemned errors ("Divine revelation is imperfect").

CHAPTER 5, IS THERE INTELLIGENT
LIFE ON EARTH?

Carl Sagan, W. R. Thompson, Robert Carlsson, Donald Gurnett, and Charles Hord, "A Search for Life on Earth from the *Galileo* Spacecraft," *Nature,* vol. 365 (1993), pp. 715-721.

CHAPTER 7, AMONG THE MOONS OF SATURN

Jonathan Lunine, "Does Titan Have Oceans?," *American Scientist,* vol. 82 (1994), pp. 134-144.

Carl Sagan, W. Reid Thompson, and Bishun N. Khare, "Titan: A Laboratory for Prebiological Organic Chemistry," *Accounts of Chemical Research,* vol. 25 (1992), pp. 286-292.

J. William Schopf, *Major Events in the History of Life* (Boston: Jones and Bartlett, 1992).

CHAPTER 8, THE FIRST NEW PLANET

I. Bernard Cohen, "G. D. Cassini and the Number of the Planets," in *Nature, Experiment and the Sciences,* Trevor Levere and W. R. Shea, editors (Dordrecht: Kluwer, 1990).

CHAPTER 9, AN AMERICAN SHIP AT THE
FRONTIERS OF THE SOLAR SYSTEM

Murmurs of Earth, CD-ROM of the *Voyager* interstellar record, with introduction by Carl Sagan and Ann Druyan (Los Angeles: Warner New Media, 1992), WNM 14022.

Alexander Wolszczan, "Confirmation of Earth-Mass Planets Orbiting the Millisecond Pulsar PSR B1257+12," *Science,* vol. 264 (1994), pp. 538-542.

CHAPTER 12, THE GROUND MELTS

Peter Cattermole, *Venus: The Geological Survey* (Baltimore: Johns Hopkins University Press, 1994).

Peter Francis, *Volcanoes: A Planetary Perspective* (Oxford: Oxford University Press, 1993).

CHAPTER 13, THE GIFT OF *APOLLO*

Andrew Chaikin, *A Man on the Moon* (New York: Viking, 1994).

Michael Collins, *Liftoff* (New York: Grove Press, 1988).

Daniel Deudney, "Forging Missiles into Spaceships," *World Policy Journal,* vol. 2, no. 2 (Spring 1985), pp. 271-303.

Harry Hurt, *For All Mankind* (New York: Atlantic Monthly Press, 1988).

Richard S. Lewis, *The Voyages of Apollo: The Exploration of the Moon* (New York: Quadrangle, 1974).

Walter A. McDougall, *The Heavens and the Earth: A Political History of the Space Age* (New York: Basic Books, 1985).

Alan Shepherd, Deke Slayton et al., *Moonshot* (Atlanta: Hyperion, 1994).

Don E. Wilhelms, *To a Rocky Moon: A Geologist's History of Lunar Exploration* (Tucson: University of Arizona Press, 1993).

CHAPTER 14, EXPLORING OTHER WORLDS
AND PROTECTING THIS ONE

Kevin W. Kelley, editor, *The Home Planet* (Reading, MA: Addison-Wesley, 1988).

Carl Sagan and Richard Turco, *A Path Where No Man Thought: Nuclear Winter and the End of the Arms Race* (New York: Random House, 1990).

Richard Turco, *Earth Under Siege: Air Pollution and Global Change,* (New York: Oxford University Press, in press).

CHAPTER 15, THE GATES OF THE
WONDER WORLD OPEN

Victor R. Baker, *The Channels of Mars* (Austin: University of Texas Press, 1982).

Michael H. Carr, *The Surface of Mars* (New Haven: Yale University Press, 1981).

H. H. Kieffer, B. M. Jakosky, C. W. Snyder, and M. S. Matthews, editors, *Mars* (Tucson: University of Arizona Press, 1992).

John Noble Wilford, *Mars Beckons: The Mysteries, the Challenges, the Expectations of Our Next Great Adventure in Space* (New York: Knopf, 1990).

CHAPTER 18, THE MARSH OF CAMARINA

Clark R. Chapman and David Morrison, "Impacts on the Earth by Asteroids and Comets: Assessing the Hazard," *Nature,* vol. 367 (1994), pp. 33-40.

A. W. Harris, G. Canavan, C. Sagan, and S. J. Ostro, "The Deflection Dilemma: Use vs. Misuse of Technologies for Avoiding Interplanetary Collision Hazards," in *Hazards Due to Asteroids and Comets,* T. Gehrels, editor (Tucson: University of Arizona Press, 1994)

John S. Lewis and Ruth A. Lewis, *Space Resources: Breaking the Bonds of Earth* (New York: Columbia University Press, 1987).

C. Sagan and S. J. Ostro, "Long-Range Consequences of Interplanetary Collision Hazards," *Issues in Science and Technology* (Summer 1994), pp 67–72.

CHAPTER 19, REMAKING THE PLANETS

J. D. Bernal, *The World, the Flesh, and the Devil* (Bloomington, IN: Indiana University Press, 1969; first edition, 1929).

James B. Pollack and Carl Sagan, "Planetary Engineering," in J. Lewis and M. Matthews, editors, *Near-Earth Resources* (Tucson: University of Arizona Press, 1992).

CHAPTER 20, DARKNESS

Frank Drake and Dava Sobel, *Is Anyone Out There?* (New York: Delacorte, 1992).

Paul Horowitz and Carl Sagan, "Project META: A Five-Year All-Sky Narrowband Radio Search for Extraterrestrial Intelligence," *Astrophysical Journal,* vol. 415 (1992), pp. 218-235.

Thomas R. McDonough, *The Search for Extraterrestrial Intelligence* (New York: John Wiley and Sons, 1987).

Carl Sagan, *Contact: A Novel* (New York: Simon and Schuster, 1985).

CHAPTER 21, TO THE SKY!

J. Richard Gott III, "Implications of the Copernican Principle for Our Future Prospects," *Nature,* vol. 263 (1993), pp. 315-319.

CHAPTER 22, TIPTOEING
THROUGH THE MILKY WAY

I. A. Crawford, "Interstellar Travel: A Review for Astronomers," *Quarterly Journal of the Royal Astronomical Society,* vol. 31 (1990), p. 377.

I. A. Crawford, "Space, World Government, and 'The End of History,' " *Journal of the British Interplanetary Society,* vol. 46 (1993), pp. 415–420.

Freeman J. Dyson, *The World, the Flesh, and the Devil* (London: Birkbeck College, 1972).

Ben R. Finney and Eric M. Jones, editors, *Interstellar Migration and the Human Experience* (Berkeley: University of California Press, 1985).

Francis Fukuyama, *The End of History and the Last Man* (New York: The Free Press, 1992).

Charles Lindholm, *Charisma* (Oxford: Blackwell, 1990). The comment on the need for a *telos* is in this book.

Eugene F. Mallove and Gregory L. Matloff, *The Starflight Handbook* (New York: John Wiley and Sons, 1989).

Carl Sagan and Ann Druyan, *Comet* (New York: Random House, 1985).

ACKNOWLEDGMENTS

Most of the material in this book is new. A number of chapters have evolved from articles first published in *Parade* magazine, a supplement to the Sunday editions of American newspapers which, with an estimated 80 million readers, may be the most widely read magazine in the world. I am greatly indebted to Walter Anderson, the editor-in-chief, and David Currier, the executive editor, for their encouragement and editorial wisdom; and to the readers of *Parade,* whose letters have helped me understand where I have been clear, and where obscure, and how my arguments are received. Portions of other chapters have emerged from articles published in *Issues in Science and Technology, Discover, The Planetary Report, Scientific American,* and *Popular Mechanics.*

Aspects of this book have been discussed with a large number of friends and colleagues, whose comments have greatly improved it. Although there are too many to list by name, I would like to express my real gratitude to all of them. I want especially, though, to thank Norman Augustine, Roger Bonnet, Freeman Dyson, Louis Friedman, Everett Gibson, Daniel Goldin, J. Richard Gott III, Andrei Linde, Jon Lomberg, David Morrison, Roald Sagdeev, Steven Soter, Kip Thorne, and Frederick Turner for their comments on all or part of the manuscript; Seth Kaufmann, Peter Thomas, and Joshua Grinspoon for their help with tables and graphs; and a brilliant array of astronomical artists, acknowledged at each illustration, who have permitted me to showcase some of their work. Through the generosity of Kathy Hoyt, Al McEwen, and Larry Soderblom, I've been able to display some of the exceptional photomosiacs, airbrush maps, and other reductions of NASA images accomplished at the Branch of Astrogeology, U.S. Geological Survey.

I am indebted to Andrea Barnett, Laurel Parker, Jennifer Bland, Loren Mooney, Karenn Gobrecht, Deborah Pearlstein, and the late Eleanor York for their able technical assistance; and to Harry Evans, Walter Weintz, Ann Godoff, Kathy Rosenbloom, Andy Carpenter, Martha Schwartz, and Alan MacRobert on the production end. Beth Tondreau is responsible for much of the design elegance on these pages.

On matters of space policy, I have benefited from discussions with other members of the board of directors of The Planetary Society, especially Bruce Murray, Louis Friedman, Norman Augustine, Joe Ryan, and the late Thomas O. Paine. Devoted to the exploration of the Solar System, the search for extraterrestrial life, and international missions by humans to other worlds, it is the organization that most nearly embodies the perspective of the present book. Those readers interested in more information on this nonprofit organization, the largest space-interest group on Earth, may contact:

THE PLANETARY SOCIETY
65 N. Catalina Avenue
Pasadena, CA 91106
Tel.: 1-800-9 WORLDS

As is true of every book I've written since 1977, I am more grateful than I can say to Ann Druyan for searching criticism and fundamental contributions both on content and style. In the vastness of space and the immensity of time, it is still my joy to share a planet and an epoch with Annie.

INDEX

ABOUT THE AUTHOR

CARL SAGAN has played a leading role in the American space program since its inception. A consultant and adviser to NASA since the 1950s, he briefed the *Apollo* astronauts before their flights to the Moon, and was an experimenter on the *Mariner, Viking, Voyager,* and *Galileo* expeditions to the planets, including the first successful planetary mission, *Mariner 2.* He helped solve the mysteries of the high temperature of Venus (answer: a massive greenhouse effect), the seasonal changes on Mars (answer: windblown dust), and the red haze of Titan (answer: complex organic molecules). He has been a pioneer in understanding the global consequences of nuclear war, in the space-craft search for life on other planets, in the hunt for radio signals from distant civilizations in space, and in laboratory studies of the steps leading to the origin of life.

For his work, Dr. Sagan has received the NASA Medals for Exceptional Scientific Achievement and (twice) for Distinguished Public Service, as well as the NASA *Apollo* Achievement Award. Asteroid 2709 Sagan is named after him. He has also been given the John F. Kennedy Astronautics Award of the American Astronautical Society, the Explorers Club 75th Anniversary Award, the Tsiolkovsky Medal of the Soviet Cosmonautics Federation, the Masursky Award of the American Astronomical Society ("for his extraordinary contributions to the development of planetary science"), and in 1994 the Public Welfare Medal, the highest award of the National Academy of Sciences (for "distinguished contributions in the application of science to the public welfare . . . No one has ever succeeded in conveying the wonder, excitement and joy of science as widely as Carl Sagan and few as well . . . His ability to capture the imagination of millions and to explain difficult concepts in understandable terms is a magnificent achievement.")

A Pulitzer Prize winner, Dr. Sagan is the author of many best-sellers, including *Cosmos* which became the most widely read science book ever published in the English language. The accompanying Emmy and Peabody award–winning television series has now been seen by 500 million people in 60 countries. He is currently the David Duncan Professor of Astronomy and Space Sciences and Director of the Laboratory for Planetary Studies at Cornell University; Distinguished Visiting Scientist at the Jet Propulsion Laboratory, California Institute of Technology; and co-founder and President of The Planetary Society, the largest space-interest group in the world.